The COVID–19 Pandemic and the Digitalization of Diplomacy

Floribert Patrick Calvain Endong
University of Dschang, Cameroon

A volume in the Advances in
Electronic Government, Digital
Divide, and Regional Development
(AEGDDRD) Book Series

Published in the United States of America by
    IGI Global
    Information Science Reference (an imprint of IGI Global)
    701 E. Chocolate Avenue
    Hershey PA, USA 17033
    Tel: 717-533-8845
    Fax: 717-533-8661
    E-mail: cust@igi-global.com
    Web site: http://www.igi-global.com

Library of Congress Cataloging-in-Publication Data

Names: Endong, Floribert Patrick C., 1983- editor.
Title: The COVID-19 pandemic and the digitalization of diplomacy / edited
  by Floribert Patrick Calvain Endong.
Description: Hershey, PA : Information Science Reference, 2023. | Includes
  bibliographical references and index. | Summary: "The COVID-19 Pandemic
  and the Digitalization of Diplomacy explores the influences of the new
  ICTs, AI, and smart cultures on the conduct of public diplomacy. It
  further examines the impact of the COVID-19 pandemic on the conduct of
  digital diplomacy in the world and analyzes the implications of the
  dynamics of ICTs and AI for teaching and research in digital diplomacy.
  Covering topics such as defense diplomacy, the fourth industrial
  revolution, and technological determinism, this premier reference source
  is an essential resource for diplomats, politicians, government
  officials, ICT developers, students and educators of higher education,
  librarians, researchers, and academicians"-- Provided by publisher.
Identifiers: LCCN 2022058458 (print) | LCCN 2022058459 (ebook) | ISBN
  9781799883944 (hardcover) | ISBN 9781799883951 (paperback) | ISBN
  9781799883968 (ebook)
Subjects: LCSH: Africa--Foreign relations--21st century. | Internet and
  international relations--Africa. | Information technology in
  international relations. | COVID-19 Pandemic, 2020---Political
  aspects--Africa.
Classification: LCC JZ1773 .C69 2023 (print) | LCC JZ1773 (ebook) | DDC
  303.48/57096--dc23/eng/20230109
LC record available at https://lccn.loc.gov/2022058458
LC ebook record available at https://lccn.loc.gov/2022058459

This book is published in the IGI Global book series Advances in Electronic Government, Digital Divide, and Regional Development (AEGDDRD) (ISSN: 2326-9103; eISSN: 2326-9111)

British Cataloguing in Publication Data
A Cataloguing in Publication record for this book is available from the British Library.

All work contributed to this book is new, previously-unpublished material.
The views expressed in this book are those of the authors, but not necessarily of the publisher.

For electronic access to this publication, please contact: eresources@igi-global.com.

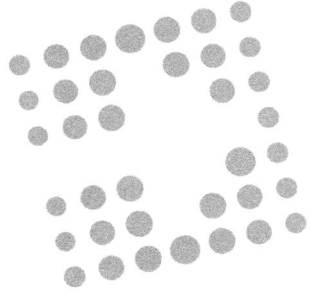

Advances in Electronic Government, Digital Divide, and Regional Development (AEGDDRD) Book Series

Zaigham Mahmood
University of Derby, UK & North West
University, South Africa

ISSN:2326-9103
EISSN:2326-9111

MISSION

The successful use of digital technologies (including social media and mobile technologies) to provide public services and foster economic development has become an objective for governments around the world. The development towards electronic government (or e-government) not only affects the efficiency and effectiveness of public services, but also has the potential to transform the nature of government interactions with its citizens. Current research and practice on the adoption of electronic/digital government and the implementation in organizations around the world aims to emphasize the extensiveness of this growing field.

    The Advances in Electronic Government, Digital Divide & Regional Development (AEGDDRD) book series aims to publish authored, edited and case books encompassing the current and innovative research and practice discussing all aspects of electronic government development, implementation and adoption as well the effective use of the emerging technologies (including social media and mobile technologies) for a more effective electronic governance (or e-governance).

COVERAGE

- Case Studies and Practical Approaches to E-Government and E-Governance
- Electronic Government, Digital Democracy, Digital Government
- E-Citizenship, Inclusive Government, Connected Government
- Knowledge Divide, Digital Divide
- Citizens Participation and Adoption of E-Government Provision
- Emerging Technologies within the Public Sector
- Current Research and Emerging Trends in E-Government Development
- Issues and Challenges in E-Government Adoption
- Online Government, E-Government, M-Government
- Urban Development, Urban Economy

IGI Global is currently accepting manuscripts for publication within this series. To submit a proposal for a volume in this series, please contact our Acquisition Editors at Acquisitions@igi-global.com or visit: http://www.igi-global.com/publish/.

Titles in this Series

For a list of additional titles in this series, please visit: http://www.igi-global.com/book-series/

Glocal Policy and Strategies for Blockchain Building Ecosystems and Sustainability
Gülsün Kurubacak (Anadolu Üniversitesi Açıköğretim Fakültesi, Turkey) Ramesh Chander Sharma (Dr. B. R. Ambedkar University Delhi, India) and Hakan Yıldırım (Purdue University, Turkey)
Information Science Reference • © 2023 • 335pp • H/C (ISBN: 9781668441534) • US $270.00

Smart Village Infrastructure and Sustainable Rural Communities
Mohammad Ayoub Khan (University of Bisha, Saudi Arabia) Bhumika Gupta (Govind Ballabh Pant Institute of Engineering & Technology, Pauri Garhwal, India) Agya Ram Verma (Govind Ballabh Pant Institute of Engineering & Technology, Pauri Garhwal, India) Pushkar Praveen (Govind Ballabh Pant Institute of Engineering and Technology, India) and Cathryn J. Peoples (Ulster University, UK)
Engineering Science Reference • © 2023 • 315pp • H/C (ISBN: 9781668464182) • US $250.00

Handbook of Research on Cyber Approaches to Public Administration and Social Policy
Fahri Özsungur (Mersin University, Turkey)
Information Science Reference • © 2022 • 692pp • H/C (ISBN: 9781668433805) • US $315.00

Advances in Deep Learning Applications for Smart Cities
Rajeev Kumar (Teerthanker Mahaveer University, India) and Rakesh Kumar Dwivedi (Teerthanker Mahaveer University, India)
Engineering Science Reference • © 2022 • 335pp • H/C (ISBN: 9781799897101) • US $250.00

Regional Economic Integration and Global Competition in the Post-COVID-19 Era
European Union, Eurasian Economic Union, and the Belt and Road Initiative
Oxana Karnaukhova (Southern Federal University, Russia) and Dmitry Shevchenko (Southern Federal University, Russia)
Information Science Reference • © 2022 • 274pp • H/C (ISBN: 9781799892540) • US $250.00

701 East Chocolate Avenue, Hershey, PA 17033, USA
Tel: 717-533-8845 x100 • Fax: 717-533-8661
E-Mail: cust@igi-global.com • www.igi-global.com

Table of Contents

Detailed Table of Contents

Chapter 1

    Fulufhelo Oscar Makananise, University of South Africa, South Africa
    Ndivhuwo Doctor Sundani, University of South Africa, South Africa

This chapter explores the contemporary trends and debates on the use of digital media
for diplomatic services and practices in the Fourth Industrial Revolution, also known
as 4IR, in the global south. The chapter emphases how digital diplomacy industrialized
and evolved to be predominant during the outbreak of the coronavirus pandemic
as known as the COVID-19 pandemic. The chapter has been strengthened by the
practice theory for digital diplomacy transformations. Most prominently, the chapter
establishes that digital media platforms played a vital role in diplomatic relations
practices for the development and social change in the global south countries during
the COVID-19 pandemic. Furthermore, the chapter suggests strategic solutions to
prevent digital diplomacy challenges and highlights the future of digital diplomatic
practices in the global south.

Chapter 2

    Floribert Patrick C. Endong, University of Dschang, Cameroon

Although a galaxy of research works have focused on diplomacy in the COVID-19
period, very little attention has particularly been given to the idea of comparing
the state of African digital diplomacy in the pre and the post COVID-19 period.
Also, the literature available suggests that most studies on digital diplomacy in the
COVID-19 period are Eurocentric and the COVID-19 pandemic as an opportunity

for the future growth of digital diplomacy in Africa is still virtually understudied. In view of filling the above-mentioned gap in knowledge, this chapter deploys secondary sources and observations to attain three principal objectives. First, the chapter explores the state of digital diplomacy before the outbreak of the COVID-19 pandemic. Second, it examines how the pandemic positively affected the conduct of digital diplomacy in Africa, putting to question the myth of digital divide between the West and Africa. Lastly, the chapter explores implications for the practice of public diplomacy in post-COVID Africa as well as for future research on African digital diplomacy.

Chapter 3

Ranson Sifiso Gwala, University of KwaZulu-Natal, South Africa
Pfano Mashau, University of KwaZulu-Natal, South Africa

The challenges brought by COVID-19 challenged all sectors of society, including public diplomacy. Public diplomacy through the adoption of social media and other digital platforms was transformed into mainstream digital diplomacy. The chapter sought to understand how social media use by diplomats during COVID-19 forced or allowed the introduction of digital diplomacy as a necessary intervention. Using the desktop review of literature, adjoining on systematic literature review was adopted as the research method. The objectives were to investigate how the outbreak of the COVID-19 pandemic affected digital diplomacy in Africa, and what are the implications of the COVID-19 pandemic for the practice of digital diplomacy in Africa. The research identified four themes, these were: the state of African digital diplomacy; social media adoption during COVID-19 in Africa; corona virus and vaccine diplomacy; and anti-corruption public digital diplomacy. The study revealed that African digital diplomacy, however, is closely comparable to its developed countries counterparts.

Chapter 4

Ekrem Yaşar Akçay, Hakkari University, Turkey
Murat Mutlu, Hakkari University, Turkey

This chapter will discuss the changes and developments in defense diplomacy in Türkiye. Türkiye, which has taken very important steps in digitalization with the COVID-19 process, has also taken important steps in defense and digitalization of defense. The study consists of two chapters. In the first chapter, the concept of defense diplomacy will be discussed and the transformation of defense diplomacy in the historical process will be examined. In the second part, defense diplomacy practices in Türkiye and the digital transformation of defense diplomacy will be examined with examples. A descriptive method will be used in the study.

Chapter 5

Chinonso Aniagu, Independent Researcher, Nigeria

Over the last decade, new media penetrations have multiplied the determinants of perceptions held in diplomatic sphere. The deployment of digital diplomacy by different countries to manage public perceptions has now opened a new vista in diplomatic and public relations discourse. China's application of digital diplomacy became even more pronounced as a result of the negative narratives--mostly derived from preconceived stereotypes--that heralded the outbreak of coronavirus. Consequently, the country's online communications in Nigeria were designed to launder its image for a more beneficial bilateral relationship with the latter. This chapter, through qualitative content analysis, evaluates the extent China's online coronavirus-oriented communications in Nigeria were deployed in countering negative stereotypes and narratives at the initial-to-peak stage of the virus. Findings show that, despite trickles of counterproductive content, post-coronavirus China's image was significantly laundered through the deployment of various Nigeria-targeted digital communication strategies.

Chapter 6

Floribert Patrick C. Endong, University of Dschang, Cameroon
Eugenie Grace Essoh Ndobo, University of Calabar, Nigeria

In a bid to mitigate the COVID-19 pandemic in its territory, the Chinese government embarked on a number of muscular policies right from the early stages of the pandemic. One of such policies was aimed at forcing Africans living in the Guangdong province to accept COVID-19 prevention and containment measures. Subsequently viewed as xenophobic, this policy rapidly degenerated into a diplomatic incident, opposing Chinese and African governments. Through its officials and its diplomats, the Nigerian government in particular condemned the Chinese policies using Twitter among other digital platforms. This chapter seeks to show how Nigerian diplomats deployed Twitter to critique China's perceived xenophobic treatment of Nigerian diasporas in its territory during the early stage of the COVID-19 pandemic. The chapter is based on a quantitative and qualitative content analysis of 12 randomly selected tweets generated by senior staffers at Nigeria's Ministry of Foreign Affairs to denounce the maltreatment of the Nigerian Diaspora in China.

    Hameed Khan, Guru Ramdas Khalsa Institute of Science and
        Technology, Jabalpur, India
    Kamal Kumar Kushwah, Jabalpur Engineering College, India

Epidemiology and scale management has been an ongoing project of governments worldwide, including India, focusing on reducing the spread of the virus and reducing the social and financial damage caused by the virus. With the closure of health infrastructure and outbreaks affecting health care workers, there is a need for a cooperative approach to the management of Covid-19 and greater involvement of public enterprises to lead government efforts to control the epidemic. The latest digital technology plays an essential role in monitoring the situation closely, assisting the government, and utilizing high-risk public organizations. Diplomacy—whether regular or non-digital—is affected: seminars and seminars are transmitted to Zoom and various video platforms; governments and international organizations have worked closely with social media, digital technology, and content boards to combat the information that is not in line with COVID-19; foreign policy actors in public and private sectors who have tried to meet new audiences, each online and offline.

    Floribert Patrick C. Endong, University of Dschang, Cameroon

The COVID-19 pandemic brought to the fore not only the centrality of Western digital technologies, but also a number of philosophico-cultural issues. Two of such cultural issues in the domain of digital diplomacy have been the de-Africanisation and dehumanisation of African digital diplomacy. These two issues have partly stemmed from the popular African myth that cybercultures in general and the digitalisation of African diplomacy in particular are disruptive forces that could negatively affect the traditional African values that have since independence been upheld by African governments in their conduct of public diplomacy. A related theory states that digitalisation may only de-humanise and de-Africanise public diplomacy. Using secondary sources and critical observations, this chapter examines the extent to which the above-mentioned fears are justified. The chapter specifically explores how digitalisation could affect specific African traditional values. It also examines the extent to which digitisation is susceptible to de-humanise or/and de-Africanise African public diplomacy.

Chapter 9

    M. Aaschita Reddy, Sreenidhi Institute of Science and Technology, India
    B Manidweep Reddy, Sreenidhi Institute of Science and Technology, India
    C. S. Mukund, Sreenidhi Institute of Science and Technology, India
    Kiran Venneti, Aditya College of Engineering, India
    D. M. D. Preethi, PSNA College of Engineering and Technology
      (Autonomous), India
    Sampath Boopathi, Muthayammal Engineering College, India

The agenda of this book chapter is to review existing technologies that aid societal health protection and recommend some possible approaches which will assist the mentioned scenario. Automations like big data and artificial intelligence (AI) deployed in healthcare sector can expedite pandemic response in ways that are strenuous to achieve all in all by humans. The sudden epiphany to trace COVID-19 in public has powered the innovation of data dashboards that visually unveil coronavirus epicentres. A cloud-based AI-assisted CT service is being engaged to differentiate pneumonia from the pandemic which dwindles risk factor in the present school of thought of the citizens worldwide. In conclusion, social health protection was an indispensable mechanism in prior to these challenging times and is escalating by prominence for delivering support to individuals during the crisis.

Preface

In a 2013 commentary article published on the *DipNote¹* blog, the former US Secretary of State, John Kerry, somewhat criticised the popular tendency among diplomacy practitioners, scholars, and critics of making the term "digital diplomacy" the talk of the town. He claimed that this term ("digital diplomacy") "is almost redundant - it's just diplomacy, period". Kerry further stressed that, although the new information and communication technologies (ICTs) do tremendously contribute to the advancement of countries' foreign policy objectives as well as to bridging the gap between people across the globe, they (the ICTs) fulfil the same core diplomatic functions as the traditional/analogue tools of public diplomacy. For instance, they enable diplomats to create dialogue among the broadest possible audience as well as to find common ground, which, after all, are what diplomacy is all about (Kerry, 2013).

For many observers, Kerry's pronouncement came to mean that it is futile to always stress the "digital aspect" of ICT-driven diplomacy, given the fact that a plurality of factors or indicators suggest that, in a near future, the use of digital technologies in diplomacy will likely become too banal that professionals and scholars in the discipline will no longer see the need to stress these "digital nature" of digital diplomacy. In spite of its pertinence, the above futuristic statement seems not to take into account a number of new developments in the domains of artificial intelligence, robotics, smart cultures, and diplomacy itself, among others. In effect, digital diplomacy itself has over the years been extremely dynamic, so much so that it is becoming increasingly complex to define players in the diplomatic game. For instance, technological innovations in AI and robotics have caused governments to fantasise over using robots as diplomats or using artificial intelligence in the conduct of consular affairs, crisis communication, public diplomacy and international negotiations. China is a good example of countries who, in recent times, have ardently resorted to AI in international negotiations and crisis communications (Daxue Consulting 2020).

In view of new developments in digital diplomacy, Bjola (2018) observes that digitalisation may not have changed the main targets of diplomacy but the truth remains that it has so transformed the diplomatic game that it will not be out of place to talk of a revolution in the conduct of diplomacy. According to Bjola, although "the core mission of diplomacy in the Digital Age is still about finding the middle ground", a lot of emergent digitally driven trend have brought significant changes in the way the diplomatic game is now conducted. What has concretely changed is "the context in which the core mission of diplomacy is supposed to be accomplished". Indeed, "new digital technologies significantly broaden the spectrum of actors that can take part and influence the diplomatic conversation, reshape the 'grammar rules' and institutional norms to guide online diplomatic engagement, and open the door to the use of digital tools for disrupting the middle ground via disinformation and propaganda" (p.8).

Besides broadening the spectrum of actors that can participate in, and shape the diplomacy game, the application of new technological innovations (notably AI and robots) in the conduct of most human industries has given birth to such paradigms as robotisation, increased and advanced mechanisation and dehumanisation of various sectors of human enterprises including public diplomacy in the world. Other related developments have been the acceleration and growing popularisation of the smart city concept as well as the COVID-19 pandemic which have all combined to compel almost all major human industries – including diplomacy – to mainly shift online and to be revolutionised day by day.

The dynamics and complexity of digital diplomacy as suggested in the foregoing, have not yet attracted the scholarly attention they deserve. The majority of studies devoted to the use of ICTs in the conduct of public diplomacy have not actually touched issues such as AI, smart technologies as well as the influence of the COVID-19 on the evolution or revolution of digital diplomacy in the world. Previous works have also not sought to address emerging issues related to the implication of the above mentioned emerging trends for teaching and research in the domain of public diplomacy. This book ultimately seeks to fill this apparent gap in knowledge. The principal objectives of the present edited volume are therefore to 1) explore the influences of the new ICTs, AI and smart cultures on the conduct of public diplomacy; 2) examine the impact of the COVID-19 pandemic on the conduct of digital diplomacy in the world, and 3) analyse the implications of the dynamics of ICTs and AI for teaching and research in digital diplomacy.

A NEW PANDEMIC FUELS A NEW NORMAL

By the time the World Health Organisation (WHO) declared COVID-19 a pandemic, governments and human industries across the world had already come to admit the need to look towards a new normal. Definable as a situation in which world economies and societies adopts new patterns of life and production following the outbreak of a crisis, this new normal was just imminent. Previous global crises such as the First World War, the September 11 attacks, the 2008-2009 financial crisis and the recent 2012-2013 global recession had in contemporary history, individually given birth to a new normal (Ghose, 2020). It was thus, not surprising that the COVID-19 in its own turn leads to a serious global change in human behaviours and a transformation of almost all human industries.

The growing mortality rates and the imperative to curb the propagation of the pandemic in the world motivated the adoption of both national and international preventive measures that were just bound to transform all sectors of human life on earth. The globally popular policies that have to some extent affected human life have been social distancing, regional quarantine (lockdowns) and vaccinations among others. In line with social distancing and quarantine policies, countries learned to live amidst closed borders, restricted international movements and the imperative and snobbism to be vaccinated. One of the most talked about aspects of the COVID-related new normal is the human societies' ability to live and thrive in spite of the continuous propagation and persistence of the Corona virus. The President of the International Federation of the Red Cross and Red Crescent Societies (IFRC) indirectly underlines this truism thus: "Living with the virus' is a privilege that many countries and communities around the world cannot enjoy. Ensuring equitable access to vaccines, diagnostics and treatments will not only save lives, but will also protect the world against the emergence of new and more dangerous variants. It is the only path to normalcy. None of us is safe until we all are" (cited in IFRC, 2022, p.6).

Another key aspect of the new normal has been the accelerated digitalisation of almost all human industries. The popular and spectacularly naturalised policies of social distancing and quarantine systematically intensified digital cultures. Indeed, from e-commerce through e-administration to e-learning and e-banking, most human enterprises/industries migrated online. All conditions were therefore made favourable for digital diplomacy – particularly in countries with reduced internet penetration – to be accelerated and regarded as the first resort.

THE DIPLOMATIC WORLD GETS ABREAST OF THE NEW NORMAL

The COVID-19 pandemic made it imperative for most human industries/ businesses to migrate online, thereby bringing to the fore the indispensability of such concepts as digitalisation, the "wired society" and artificial intelligence among other new technologies. However, digital diplomacy – like many other cybercultures – is neither new to the world's diplomats and MFAs; nor is it a revolution that emerged during, or as a result of the COVID 19. This digital culture is as old as the Internet and telecommunications themselves and has been a very popular concept particularly in the West long before the advent of the COVID-19 pandemic. The Global South may have lagged behind the West in terms of the application of ICTs in the conduct of public diplomacy; however, the fact remains that digital diplomacy is a pre-COVID innovation which difficult circumstances made very popular and indispensable. History has it that, digital diplomacy emerged in the West well before the 2000s (Metzger, 2012; Roberts, 2007). According to Kurbalija (2022), the very first online participation session in multilateral diplomacy took place in 1963. This online session was organised by the International Telecommunication Union (ITU). Since then, the availability of the internet in conference room has enabled remote participation in contexts of international negotiations.

In spite of its age-old nature, e-diplomacy has over the years, suffered from a somewhat inferior status, compared to traditional analogue diplomacy. As noted by a good number of observers, before the outbreak of the COVID-19 pandemic, issues such as online meetings were popularly considered as poor substitutes for face-to-face ones (Diplo, 2020; Hayden, 2012; Cull, 2008; Dizrad Jr., 2001). For some observers (notably former US Vice-President John Kerry), digital diplomacy didn't deserve much publicity as it was diplomacy in the same right as traditional/offline diplomacy. In the same line of thought, many other observers, regarded e-diplomacy approaches such as online summits as technologies, which, compared to traditional diplomatic tools, offered less opportunities for effective negotiations. Adesina (2022) explains this truism thus "physical meetings are important because negotiations often take place on the sidelines of international gatherings, through impromptu discussions between global leaders at tea or lunch breaks, or a chance encounter in the corridor or rest room and so on. Also, physical meetings provide an opportunity for participants to observe and interpret body language and emotions of the parties, which may help in decision making" (p.95). Following the above logic, Adesina (2022) thinks that, in spite of its successes, digital diplomacy can never replace traditional diplomacy.

The above-mentioned reductionist views of digital diplomacy however changed in a drastic manner with the advent of the COVID-19 pandemic. In other words, the pandemic came to change popular perceptions about digital diplomacy in the world. As rightly observed by Diplo (2020), the COVID-19 enabled a re-appreciation of the importance of digital diplomacy: the question is no longer whether diplomats should use e-diplomacy despite their drawbacks, but rather how these form of diplomacy can be used effectively and how it can be blended with the traditional one. If a number of studies do admit that the COVID-19 pandemic impacted the evolution of digital diplomacy, only few research works try to clearly theorise on the ways in which the pandemic changed the practice of digital diplomacy, particularly in the West (Labott, 2022; Bjola & Manor, 2022; Tara, 2022; Peru, 2022; UNO, 2021; Hocking and Melissen, 2015; Shenhav, Sheaer & Gabay, 2010). Few studies also focus on the Global South experience to show how the pandemic contributed in changing this region from low to a high spot of digital diplomacy (Endong 2021, 2022; Adesina 2022). Rare are studies that seek to particularly show the COVID-provoked evolution of digital diplomacy in the West where digitalisation has long been a popular culture.

This book therefore seeks to contribute to a greater understanding the impact of the COVID-19 pandemic on the status of digital diplomacy. The various chapters constituting the book focus exclusively on both western countries – where digital diplomacy has long been popular, even before the outbreak of the Corona virus - and African countries which, under the pressure born from the COVID-19, have been compelled to ameliorate their appropriation of digital diplomacy.

STRUCTURE OF THE BOOK

This book is organised in nine chapters:

Chapter 1 titled "Digital Media and their Implications on Diplomatic Practices in the Fourth Industrial Revolution: A Global South Perspective" explores the contemporary trends and debates that arise from the digitalisation of diplomacy in the global south. Based on a critical review of secondary sources, its two authors (Makananise and Sundani) argue that digital media platforms played a vital role in diplomatic relations practices for the development and social change in the global south countries during the

COVID-19 pandemic. Placing much emphasis on the ways in which the COVID-19 pandemic enabled the growth of digital diplomacy in the global south, the authors explore the future of diplomatic practices in this region of the world. They also advance strategic solutions to a number of foreseeable digital diplomatic diplomacy challenges.

"Technological Determinism and African Public Diplomacy" (chap 2), by Floribert Patrick C. Endong compares the state of African digital diplomacy in the pre and the post COVID-19 periods. Using a critical exploitation of secondary sources as well as personal observations, the chapter attains three principal objectives. First, it examines the state of African digital diplomacy before the outbreak of the COVID-19 pandemic. Second, it discusses the pandemic's positive impact on the conduct of digital diplomacy in Africa; and third, it addresses implications for the practice of public diplomacy in post-COVID Africa as well as for the future of research in African digital diplomacy.

The third chapter of the book is entitled "COVID-19; Digital Diplomacy, International Relations, Social Media and Public Diplomacy." It discusses the impact of the COVID-19 pandemic on the evolution of African digital diplomacy. Such a discussion is done in the light of various ICTs and crisis related theories. Feeding on secondary data, the chapter specifically examines four issues namely the state of African digital diplomacy; the remarkable patterns of social media appropriation and use of ICTs during the COVID-19 in Africa; the Corona virus and vaccine-based diplomacy; and anti-corruption public digital diplomacy.

In chapter 4 titled "Digitalization of Public Diplomacy: Defense Diplomacy in Türkiye," Ekrem Y. Akçay and Murat Mutlu deploy a descriptive research design to discuss the role played by digitalisation in the change and development of Turkey's defence diplomacy. In the first place, the two scholars explicate the concept of defence diplomacy placing particular emphasis on the Turkish experience. In the second place, the scholars examine the ways in which digital transformations in the country have positively impacted on the conduct of Turkish defence diplomacy.

In chapter 5 titled "China's Coronavirus-Oriented Diplomacy in Nigeria: A Content Analysis of the Chinese Embassy's Online Communication," Chinonso Aniagu critically interrogates China's use of digital diplomacy to launder its image in the aftermath of series of COVID-19 related diplomatic incidents with African countries. The chapter particularly evaluates Chinese online diplomatic communication aimed at rebranding China in Nigeria during the early stages of the COVID-19 pandemic. It shows the extent to which such communication were effective in neutralising the multiple negative stereotypes of China which had emerged as a result of the outbreak of the Corona virus and China's apparent anti-African policies during the early stages of the pandemic. The paper argues that despite trickles of counterproductive content, post-Coronavirus China's image was significantly laundered through the deployment of various Nigeria-targeted digital communication strategies.

Chapter 6 entitled "Tweeting to Vindicate the "Marginalised Nigerian Diaspora in China" examines the ways in which Nigerian diplomats deployed Twitter to critique China's perceived xenophobic treatment of Nigerian diasporas in its territory during the early stage of the COVID-19 pandemic. The chapter hinges on a quantitative and qualitative content analysis of 12 randomly selected tweets generated by senior staffers at Nigeria's Ministry of Foreign Affairs to denounce the maltreatment of the Nigerian Diaspora in China. It argues that Nigerian diplomat's online engagement with Nigerian audiences on Twitter was non-dialogic for the most part. This really limited exchanges between them (Nigerian diplomats) and their audiences and hampered a number of clarifications that could have thrown more light on the diplomatic actions of both China and Nigeria during the diplomatic crisis.

Chapter 7 is titled "Using Digital Diplomacy in Context of COVID-19 Pandemic: The Indian Experience." Authored by Hameed Khan and Kamal K. Kushwah, the chapter examines the Indian experience as concerns the effects of the COVID-19 pandemic on Indian human industries. The chapter devotes parts of its attention to the prominence of digitalisation during the early stages of the pandemic as well as on how such digitalisation revolutionised the conduct of diplomacy in India. In line with this, the chapter argues that the latest digital technologies played an essential role in the Indian government's systematic efforts towards the monitoring of the crisis. They helped the government, to establish constant communication between both foreign/national audiences as well as to combat misinformation that represented a threat to governmental and international non-governmental organisations' anti-pandemic policies.

Chapter 8 titled "Dehumanising and De-Africanising Public Diplomacy" addresses two of the multiple philosophico-cultural issues that arose as a result of the centrality of Western digital technologies in the conduct of most human industries in the early stages of the COVID-19 pandemic. The two issues include the de-Africanisation and dehumanisation of African digital diplomacy. On the bases of secondary sources and critical observations, the chapter argues that digitalisation contributed in no small measure to the dehumanisation of public diplomacy. The chapter highlights specific principles of the African diplomacy that may be affected by the continuous digitalisation of the practice.

Reddy M.A. et al's "Social Health Protection during COVID Pandemic using IOT" is the last chapter of the book. It reviews existing technologies that have been relevant to health protection during various phases of the COVID-19 pandemic. The chapter also highlights some of the ways in which the pandemic inspired interstate and inter-organisation cooperation in the domain of social health protection.

Floribert Patrick Calvain Endong
University of Dschang, Cameroon

REFERENCES

Adesina, O. (2022). *Africa and the future of digital diplomacy. Foresight Africa: Technological innovations: Creating and harnessing tools for improved livelihood*. Foresight Africa.

Bjola, C. (2017). *Trends and counter-trends in digital diplomacy. Working paper no18. "Digital Diplomacy in the 21ˢᵗ Century" project*. Munich: German Ministry of Foreign Affairs.

Bjola, C. (2018). *Diplomacy in the digital age*. El Cano Institutes.

Bjola, C. (2019). *Diplomacy in the age of artificial intelligence*. The Emirate Diplomatic Academy.

Bjola, C. (2022). Digital diplomacy as world disclosure: The case of the COVID-19 pandemic. *Place Branding and Public Diplomacy*, *18*(2), 22–25. doi:10.105741254-021-00242-2

Bjola, C., & Manor, I. (2022). The rise of hybrid diplomacy: From digital adaptation to digital adoption. *International Affairs*, *98*(2), 471–491. doi:10.1093/ia/iiac005

Cull, N. J. (2008). Public diplomacy: Taxonomies and histories. *The Annals of the American Academy of Political and Social Science*, *616*(1), 31–54. doi:10.1177/0002716207311952

Daxue Consulting. (2020). *The AI ecosystem in China*. Daxue Consulting.

Diplomacy Data. (2016). History of digital diplomacy and main milestones. *Diplomacy Data*. http://diplomacydata.com/history-of-digital-diplomacy-and-main-milestones/

Dizrad, W. Jr. (2001). *Digital diplomacy: US foreign policy in the information age*. Praeger.

Endong, F. P. C. (2020). Digitisation of African public diplomacy: Issues, challenges and opportunities. *IJDS: International Journal of the Digital Society*, *11*(2), 1607–1618. doi:10.20533/ijds.2040.2570.2020.0201

Endong, F. P. C. (2022). Re-branding China's battered image in Nigeria amidst the COVID-19 pandemic. A qualitative analysis of Chinese diplomatic communications. *Journal of BRICS Studies*, *1*(1), 26–40. doi:10.36615/jbs.v1i1.615

Ghose, S. (2020). *Crisis as catalyst: The COVID-19 impact on innovation*. Berkeley. Sutardia Center or Entrepreneurship & Technology.

Hayden, C. (2012). Social media at stake: Power, practice, and conceptual limits for US public diplomacy. *Global Media*, *25*, 1–15.

Hocking, B., & Melissen, J. (2015). *Diplomacy in the digital age*. Clinhendeal Institute.

IFRC. (2020). *COVID-19 two years on: A new normal for some while millions still at risk, warns Red Cross Red Crescent*. IFRC.

Kerry, J. (2013). Digital Diplomacy: Adapting Our Diplomatic Engagement. *DipNote*. U.S Department of State Official Blog, 6/V/2013. http://2007-2017-blogs.state.gov/stories/2013/05/06/digital-diplomacyadapting-our-diplomatic-engagement.html

Kurbalija, J. (2022). Digital diplomacy. *Diplo*. https://www.diplomacy.edu/topics/digital-diplomacy/

Labott, E. (2022). *Redefining diplomacy in the wake of the COVID-19 pandemic*. The Meridian Center for Diplomatic Engagement.

Manor, I. (2018). *The digitalisation of diplomacy: Towards clarification of a fractured terminology. Working Papers Series*. Oxford: Oxford Digital Diplomacy Research Group.

Metzger, E. T. (2012). Is it the medium or the message? Social media, American public relations and Iran. *Global Media Journal*, *38*, 1–16.

Peru T. (2022). Digital diplomacy is the next normal. *Diplomatic Courier*, 20-22

Roberts, W. R. (2007). What is public diplomacy? Past practices, present conduct, possible future. *Mediterranean Quarterly*, *18*(4), 36–52. doi:10.1215/10474552-2007-025

Shenhav, S. R., Sheafer, T., & Gabay, I. (2010). Incoherent narrator: Israeli public diplomacy during the disengagement and the elections in Palestinian authority. *Israel Studies*, *15*(3), 143–162. doi:10.2979/isr.2010.15.3.143

Tara, N. (2022). *Africa needs smarter investment in digital infrastructure: Strategies for enticing the private sector. Foresight Africa: Technological innovations: Creating and harnessing tools for improved livelihood.* Foresight Africa.

United Nation Organisation. (2021). *Digital diplomacy in the era of COVID-19.* UNO.

ENDNOTE

[1]   *DipNote* is the name of the US Department of State's official blog.

Acknowledgement

I am highly indebted to all the people who were directly or indirectly involved in the development of this book, from the authors who variously interpreted the themes of the book project to the reviewers and colleagues who took the pain to provide inestimable suggestions for the amelioration of the various contents of the chapters.

First, I would like to thank the authors who are the main brains behind this book. *The COVID-19 Pandemic and the Digitalisation of Diplomacy* would not exist without their hard work, thoughtful insights, flexibility and patience.

Second, I wish to appreciate the invaluable contribution of reviewers who provided precious comments on the respective chapters as well as recommendations on how these manuscripts could be reworked and made publishable. A special thank is here addressed to Dr. Shalin Hai-Jew, Dr. Ashon C. Ashon, and Dr. Effiong Charles for their timely feedback on some of the chapter contributions. Most of the authors equally took part in the review process. Their expertise and thoughtful recommendations are greatly appreciated.

Finally, I would like to acknowledge the efforts of IGI Global staff, particularly Ms. Joselynn Hessler whose guidance greatly enabled the development of this book.

Floribert Patrick C. Endong
University of Dschang, Cameroon

Chapter 1
Digital Media and Their Implications on Diplomatic Practices in the Fourth Industrial Revolution:
A Global South Perspective

Fulufhelo Oscar Makananise
https://orcid.org/0000-0002-9360-5863
University of South Africa, South Africa

Ndivhuwo Doctor Sundani
University of South Africa, South Africa

ABSTRACT

This chapter explores the contemporary trends and debates on the use of digital media for diplomatic services and practices in the Fourth Industrial Revolution, also known as 4IR, in the global south. The chapter emphases how digital diplomacy industrialized and evolved to be predominant during the outbreak of the coronavirus pandemic as known as the COVID-19 pandemic. The chapter has been strengthened by the practice theory for digital diplomacy transformations. Most prominently, the chapter establishes that digital media platforms played a vital role in diplomatic relations practices for the development and social change in the global south countries during the COVID-19 pandemic. Furthermore, the chapter suggests strategic solutions to prevent digital diplomacy challenges and highlights the future of digital diplomatic practices in the global south.

DOI: 10.4018/978-1-7998-8394-4.ch001

INTRODUCTION AND BACKGROUND

Diplomatic practice is a conventional service offered by diplomats and government officials that serve and oblige the well-being of their respective countries. This is especially true in the global south countries, where the diplomats and government officials would be afforded the highest responsibilities and roles to establish and report on their country's accreditation in the global community. Additionally, in the international relations and political sciences fields, this is called "diplomatic practice". Diplomatic practice is the patriotic amenity given to distinct people where they are entrusted with the responsibility to represent and participate in decision-making meetings on behalf of their countries. In this instance, the diplomats need to demonstrate sincere interest and patriotic spirit in the history, culture, politics, and major public discourses of the country to which they are accredited and desist from making a judgment on the behaviour and morals of its leaders and people (Shumba, 2020 & Zsubrinzky, 2020). In addition, diplomatic practice is usually viewed as a platform in which diplomats render services to strengthen their country's relations with other countries in the global sphere. This particular amenity should be rendered by faithful, and loyal people with high ethical and moral principles and behaviour who would not dilapidate their country's relations with other countries in the global domain. However, as much as these diplomats should adhere to the ethical principles of their operation, they are furthermore, expected to possess a philosophical intuition about the strategies and tactics that could be used to assist their respective countries to achieve their targeted national Sustainable Development Goals (SDGs) at the international level. In most cases, this is achieved through their participation in diplomatic decision-making meetings, conferences, and debates to register their presence, concerns, and standpoint on diplomatic international issues and affairs which as well affect their countries.

Erstwhile to the emergence of the internet and digital media platforms, the diplomatic practice was offered through traditional, face-to-face, or interpersonal communication, where diplomats would meet physically at a designated venue. Here, they would discuss crucial issues that affect their countries such as poverty eradication, socio-economic, education, and politics among other things. However, since the rise of the internet, new media technology, and digital media, there had been a pioneering transformation and modification in the way in which diplomatic services are rendered both local, national, continental, and global. As, Kurbalija (2017) indicated that digital diplomacy focuses on the interplay between the internet and diplomacy, ranging from internet driven-changes in the environment in which diplomacy is conducted to the emergence of new topics on diplomatic agendas such as cybersecurity, and privacy, along with the use of internet tools to practice diplomacy. In addition, the integration of the use of digital media for diplomatic service resulted in what is commonly known or called "digital diplomacy", which will be discussed in detail in the subsequent section of this chapter, where diplomats use digital media platforms to address core issues and attend diplomatic strategic meetings.

Furthermore, Adesina (2017, p.1) defined digital diplomacy as "the use of digital technologies and social media platforms such as *Twitter, Facebook*, and *Weibo* by States to enter into communication with foreign publics usually in a non-costly manner". Noticeably, even though, countries in the global south such as South America, African countries, China, India, Mangolia, Saudi Arabia, and many more incorporated the use of these technologies to participate in global diplomatic and political matters; diplomatic services or practices have not been entirely compromised. This is evident in most diplomatic virtual meetings, where the various diplomats still respect and follow the basic rules, procedures, and rituals that surround their diplomatic activities and proceedings. To support the exceeding assertation, in May 2020, during the outbreak of the COVID-19 pandemic globally, the African Union also known as AU hosted a virtual conference titled "*Silencing the guns in Africa by 2020*", where diplomatic officers from diverse African countries participated in the debates, discourses, and made new connections. The AU is a continental body comprising 55 member states and diplomats from various African countries. In this occurrence, the conference was attended both physically and virtually by diplomats and government officials to address and interrogate how African countries could jointly fight or prevent conflicts, wars, genocide activities, gender-based violence epidemics, femicides, and prevent xenophobic attacks that usually unfold in many African countries and abroad.

Furthermore, the chapter emphases on the contemporary trends and debates concerning the use of digital media for diplomatic practices in the 4IR in the global south. Most prominently, it provides detailed scrutiny of the use and roles of digital media in diplomatic relations practices for development and social change during the COVID-19 pandemic in the global south. The chapter highlights the impact of the COVID-19 pandemic on digital diplomacy in the global south. In addition, this chapter analyses *how* global south countries use digital media platforms for diplomatic practices such as negotiating governance, nation-building programmes, and socio-economic expansions during the COVID-19 pandemic. It further assesses the extent to which countries in the global south use digital media technologies to address their socio-economic possible challenges, benefits, and risks involved. It further addresses various complications brought by the 4IR that affect diplomatic practices during the digital communication processes. Lastly, the chapter suggests strategic solutions to prevent digital diplomacy challenges in the global south and highlights the future of digital diplomatic practices in the global south.

LITERATURE REVIEW

Chronological Development of Digital Diplomacy in the Global South

Digital diplomacy is ordinarily regarded as the use of the internet, electronic mail, broadcast media, digital media platforms, social media sites, and all technologically advanced inventions by government officials and diplomats to achieve their diplomatic objectives, and missions (Wekesa, 2020). As a central principle of electronic, telecommunication, and diplomatic communication, digital diplomacy has its origin in diverse fields or disciplines of study. In simple terms, digital diplomacy is engrained in diverse disciplines such as Communication Science, Media Studies, International Relations, and Political Sciences which are embedded in Humanities and Social Sciences. Digital diplomacy is essentially an interaction that happens between the internet and diplomatic practices to address diplomatic objectives and various affairs. As, Ittefaq (2019) contended that digital media and new technology open up new avenues for governments, individuals, and organisations to engage with foreign audiences and communities in the global virtual space. However, developing countries governments are still deficient in the recognition of the potential of social and digital media platforms. This is about how various government officials and diplomats could effectively use digital media, platforms, the internet, online conferencing, and new media technology such as social media platforms: i.e *Twitter,* and *Facebook* for proper governance, establish power relations, and to accomplish their diplomatic missions, sustainable development goals, and persuasive objectives.

Furthermore, existing literature and research show that digital diplomacy is a phenomenon that started to be popular in the global north towards the end of the Third Industrial Revolution. This revolution is believed to have brought with it all types of massive advances and social transformations such as economics, transportation, politics, e-government, e-commerce, the internet, technology, and printing machines internationally. Auspiciously enough, it was during this period that digital diplomacy as an e-government or e-diplomatic phenomenon found its steadiness and home within various countries, and societies. In the same vein, Mohajan (2021) designated that the third industrial revolution began in the 1950s and reached its peak in the dot.com era of the late 1990s, with the movement from mechanical and analogue electronic technology to digital electronics, such as green buildings, electric cars, and trade industries. As previously indicated, this is based on energy transition, digital technologies, and the internet, which are commonly known as the "digital revolution". Recently, most people in academia started a discourse about "digital humanities" or the "digitalisation of humanities". It is during this time when societies were keen to use new media technologies, that digital diplomacy was also properly conceptualised. In this instance, one could claim that different countries in the global south strongly embraced the use of digital media for diplomatic practices and services for their benefit. This could include the operation of *Facebook* and *Twitter* accounts by a large number of African foreign ministries, thus Ethiopia's ministry of foreign affairs and embassies heavily leveraging social media for diaspora diplomacy (Turianskyi & Wekesa, 2021). Henceforth, the need to scrutinise the long-term expansion and growth of digital diplomacy in the global south communities is crucial.

Furthermore, it was during the disruptive time of the COVID-19 outbreak and the quick emergence of the fourth industrial revolution in 2020, that most people including academics and government officials begin to think about it as the start of digital diplomacy. However, this imprint needs to be addressed because just like other important sectors digital diplomacy started decades ago. As, studies conducted by Hocking and Melissen (2015), Adesina (2017), Turianskyi and Wekesa (2021) indicated that the evolution of digital diplomacy is dated back to the 1860s with the introduction of the telegraph, and subsequently the expansion of radio, television, telephones, faxes, the internet, and now new media technologies and social media such as *Twitter* and *Facebook* platforms. Ritto (2014) once specified that the invention of these media platforms over the years assisted in improving communications between diplomats, government officials, and diplomatic countries that previously relied on geographical proximity. The advance has added to the speed and precision of communications, information sharing, and participation in decision-making powers with other diplomatic countries.

In addition, the telegraph and these other broadcast inventions were first introduced and used in the First World countries or global north countries and later used or practiced in the global south countries, including South Africa. Thereafter, these media developments have influenced diplomatic practices and services. As, Manor (2019) indicated that throughout the history of diplomacy, digital media have had a deep influence and inspiration on the conventional diplomatic practices and services of various countries. In addition, these sequential expansions over the centuries show that diplomacy as a substantial phenomenon has endured through diverse dispensations of technological inventions, change, and social growth. In a nutshell, Turianskyi and Wekesa (2021) asserted that the rise in these technological developments between the 20th and 21st centuries had ushered in the digital information age and the use of digital media for diplomatic practice and services in different parts of the world, including global south countries such as South Africa. With the outline of the fourth industrial revolution, digital diplomacy continues to influence diplomatic functions, roles, and practices.

Use of Digital Media for Diplomatic Relations Practices in the Global South

Losifidis and Wheeler (2016) highlighted that a considerable number in the global south countries have recognised the prospects offered by these new technologies and have embraced them. Whereas Hocking and Melissen (2015) discovered that 'newness' in diplomacy today has everything to do with the application of new communications technologies to diplomacy. This issue goes right to the heart of diplomacy's core functions, including negotiation, representation, and communication processes. As far as the use of digital media for diplomatic relations practices is concerned, Losifidis and Wheeler (2016) held that technological developments in the field of digital communication have revolutionised the practice of public diplomacy. Whereas Bjola and Manor (2022) indicated that the past decade has seen tremendous growth among the various diplomats' in the use of digital technologies. Bjola and Manor further explained that digital media have transformed the space within which diplomats communicate, engage, and collaborate or even the logic and working procedures of diplomatic institutions, which now seek to copy those of media institutions. Commenting on the use of digital media for diplomatic relations practices in the global south, Bjola and Manor further designated that being on *Facebook,* diplomats help them to monitor online conversations in real-time and learn to anticipate possible shocks to the international system.

The study conducted by Endong (2020) revealed that till today, from Kenya and South Africa through Mali and Namibia to Rwanda, digital diplomacy has become a received idea in the African continent. Endong further indicated that a 2019 survey conducted by *Twiplomacy*, reports on efforts by specific African leaders such as Kenya's Uhuru Kenyatta, Ghana's Nana Akufo-Ado, and Rwanda's Paul Kagame towards using social media platforms such as *Twitter, Facebook,* and *Instagram* for diplomatic motives. As findings of the use of digital media for diplomatic relations practices, Melissen and de Keulenaar (2017) indicated that *Facebook* and *Twitter* are now commonly used in the corridors of diplomacy to gather information, communicate ideas, strategies and policies, build relationships, manage networks and to crowd-source knowledge. The above-mentioned discourse is supported by Wekesa (2020) who explained that this type of digital diplomacy is heavily communications-based and is undertaken through popular social networking sites such as *Twitter*, *Facebook*, and *LinkedIn*, just to name but a few.

Role of Digital Media on Diplomatic Communications in the Global South

According to Bjola and Manor (2022) since 2003 diplomatic services have experimented with establishing virtual embassies, creating social media channels to interact with foreign populations, launching smartphone applications, establishing new digital task forces, assembling big data units, revamping communication procedures in multilateral organisations and writing their procedures. In this regard, Endong (2020) indicated that the diplomacy dialogue on digital media is also more strategic than the monologue as it offers greater opportunities for interactions with foreign publics, a situation which is susceptible to facilitate the establishment of a stronger relationship between the digital diplomat and foreign publics. Additionally, Bjola and Manor (2022) are of concern that social media have had an 'empowering effect' on diaspora communities in their relationship with the diplomatic institutions of countries of origin, leading them to develop variable configurations of political, economic, and cultural engagement with the ministry of foreign affairs and embassies. As findings of the roles of digital media, Hocking and Melissen (2015) explained that networking as the conceptual basis of modern diplomatic practice including its digital dimension has fundamental implications for conceptualising and practising diplomacy, for office routines and rules of engagement among people representing different types of public and private actors, and in a more general sense for officials engaging with the outside world.

The study conducted by Bjola and Manor (2022) further revealed the fact that digital technologies enable the proper execution of diplomatic tasks and objectives. Also, these technologies alter the landscape in which diplomacy takes place or induce behavioural change speaks volumes about the multifaceted and incisive effect these technologies increasingly have on diplomatic practices and institutions. For many, the role of digital media seems to be equated with the broader public diplomacy function, with diplomats embracing *Twitter* and *Facebook*, and their embassies engaging with local audiences in the digital domain (Hocking & Melissen, 2015). Most significantly, since their inception, digital media play an essential role in social, economic, educational, political, and cultural developments and the expansion of foreign policy.

Challenges and Opportunities of Digital Diplomacy Practice in the 4ir

In his attempt to define the Fourth Industrial Revolution, Schwab (2016) designated that the digital revolution is not something new, it has been part of the public since the middle of the third century. The revolution is characterised by a fusion of technological innovations that are blurring the lines between the physical, digital, and biological spheres of life. As, McGinnis (2020) stated that the Fourth Industrial Revolution is a way of describing the blurring of boundaries between the physical, digital, and biological worlds. This shows that these digital technologies have been with us, as part of our daily communication and interaction, even in diplomatic practices and services for years. This means that the 4IR has created opportunities for government officials, and diplomats, from different parts of the world to connect through the use of the internet. For instance, in the quest to use this new media technology in May 2020, the AU launched its digital transformations strategy for the period 2020-2030, a potentially positive development that could be built on to develop a continent-wide digital diplomacy strategy in the same manner as the European Union (African Union, 2020).

In addition, Moll (2022) indicated that the Fourth Industrial Revolution or Industry 4.0 is characterised by an exponential pace of technological developments covering wide-ranging fields such as artificial intelligence (AI), robotics, the internet of things (IoT), 3D printing, information, and communication technology, materials science, energy storage, and quantum computing, to name but a few. Over the years it has been evident that one of the opportunities of the use of digital media technology is that it builds on the exponential growth of digital capacities, blurring the lines between the government's physical and digital domains. The digital media platforms in the fourth industrial revolution have overturned the diplomatic profession which is traditionally anchored on physical meetings, and interpersonal interactions among diplomats. As, Turianskyi and Wereka (2021) specified that during this time digital diplomacy was swiftly becoming embedded in the daily conduct of political, international, and foreign affairs or matters. For instance, through the use of digital platforms, diplomats and representatives from various countries could converge through online systems to debate on strategic-diplomatic issues. This meeting and interaction strategy is also in a way cost-efficient, as key diplomats and stakeholders could come together, discuss, negotiate, and take quick diplomatic decisions without paying any traveling costs, accommodation, and meals.

Impact of the Covid-19 Pandemic on Digital Dimplomacy in the Global South

Over the years, diplomacy has been one of the extant areas throughout the different revolutions including the current revolution of artificial intelligence and the internet of things. Whereas it is apparent that at the dawn of every revolution, digital media platforms challenged, disrupted, threatened, and affected the operation and practices of traditional diplomatic public services and foreign affairs globally. Turianskyi and Wereka (2021, pp. 344) indicated that "digital technologies have exerted an impact on diplomacy, and continue to do so- often in disruptive ways." Moreover, during the transition period, the diplomats and government officials have to alter the way diplomatic services are offered to meet recent technological practices, innovations, and standards. In addition, scientific investigations and research show that for several years now, both traditional and digital diplomacy has been co-existent and as such are not in competition but complement each other. As, Adesina, (2017), and Shumba (2020) highlighted even though digital media technology has successfully transmuted the way societies, communities, and even countries communicate and exchange information; the process that transformed the political, social, and economic landscape across the world; digital diplomacy which is also known as e-diplomacy, diplomacy 2.0, twiplomacy, and cyber-diplomacy should not be considered as a replacement but as complementary to traditional diplomacy practices (Faye, 2000). In return, technological innovations should be used to support the diplomats to deliver their services and articulate their diplomatic objectives, international communications, negotiations, and foreign policy affairs in the global space. For instance, since the outbreak of COVID-19 in 2020 the former African Union Chairman, President Cyril Ramaphosa of South Africa has held several online or virtual meetings and engagements through different digital media platforms with various stakeholders, parliamentarians, diplomats, and representatives of various African counties to deliberate numerous pertinent health and governance issues relating to how the African continent could fast combat the spread of the coronavirus pandemic and economic difficulties.

Furthermore, Mhlanga and Moloi (2020, pp. 2) highlighted that "as the physical, digital, and biological worlds continue to converge due to the fourth industrial revolution, new technologies and platforms will increasingly enable citizens to engage with governments and political parties to voice their opinions, coordinate their efforts, and even circumvent the supervision of public authorities". In addition, the 4IR has revealed the great extent to which the modern world both the global north and south ultimately and entirely depends on digital media platforms and technological innovations to advance their economic, educational, business, and political activities. Broadly, Turianskyi and Wereka (2021) advocated that almost all nations of the world with international and multinational organisations are now part of the digital diplomatic sphere. Digital diplomacy is more than the application of social media to familiar diplomatic functions, even though the challenge of their use in diplomacy should not be underestimated. Many diplomats and foreign ministries still apply analogue habits and norms to a digital world (Hocking & Melissen, 2015). Similar to other domains of life, diplomacy, and diplomatic officials also use digital media platforms such as Zoom, Google Meet, and Microsoft Teams to suggest their voice's interests in international policymaking decisions and free dissemination of information, whether accurate or not, relating to any issue or event. This affords diplomatic officials in the global south opportunity to participate in virtual negotiations, global decision-making gatherings, meetings, and virtual conferences.

Digital Diplomatic Practices for Development Communication and Social Change

Besides the growing penetration of the internet in the African continent and global south communities, there is a growing social media penetration supplemented by rising mobile connectivity (Endong, 2020). It was further emphasised that such growing access to the Internet and mobile has been called the "smartphone revolution" which is susceptible to facilitate not only e-cultures proliferation such as "mobile governance" (m-governance) but also citizen diplomacy, development, and social change. In addition, the study conducted by Masters (2021) on the fourth industrial revolution, and digital diplomacy revealed that the late 2000s saw a growing interest in the potential of digital technology in the global south. Especially when corporate actors, donor agencies, civil society, government stakeholders, and diplomats are increasingly implementing digital technologies and social media for development communication and social change purposes in the global south. Also, Masters (2021) is of concern that given questions around the protection and use of big data, the spread of misinformation, and the rights of citizens when it comes to the use and access of digital technologies, digital diplomacy must certainly be conceptualised to include the negotiations that shape the international digital regime.

Furthermore, Melissen and de Keulenaar (2017) are of the view that digital diplomatic practices should also be recognised as an agent of development communication and social change in most emergent countries. Bjola and Manor (2022) argued that digital technologies even 'democratise' diplomacy, and empower non-state actors and the public at large to challenge the authority of the state. Similarly to this, Endong (2020) found that for global south communities to advance and establish their diplomatic goals and objectives they have to advocate for the use of digital technology to efficiently carry out the functions of diplomats such as developing communication and social change. Bjola and Manor (2022) indicated that diplomats' ability to improve their digital experience substantially informed their views regarding the contribution that virtual platforms made to their work. Furthermore, Endong (2020) explained that these practices include representation and promotion of the home nation, establishing both bilateral and multilateral relations, consular services, and social engagement. It encapsulates the adoption of multiple ICT tools over the Internet to support a nation's interests in other countries while ensuring that foreign relations are improved between the countries.

Pertaining to engagement for social change by diplomats, Hocking and Melissen (2015) found that the digital domain for instance opens up new forms of engagement opportunities. Thus, Losifidis and Wheeler (2016) indicated that the use of Web 2.0 technologies to communicate directly with the public is interesting. Consequently, their public diplomacy 2.0 strategies have included online petitions, the hosting of campaign websites, charitable engagements, and partnerships with interested parties to affect international protests. ICTs and digital media according to Wekesa et al., (2021) have been inscribed into diplomatic practice at the outward image promotion levels and inward data and information management levels in traditional and non-traditional diplomatic organisations. In most cases, these digital media platforms can drive the countries' communication for developmental and social transformation purposes.

On the other hand, Bjola and Manor (2022) indicated that opening social media accounts, and training diplomats to use them, require much less effort and resources than establishing strategic communication and diplomatic change. Further, the above-mentioned authors clarified that digital adoption, on the other hand, involves more complex learning and communication processes. As for Endong (2020) Africa is lagging in terms of Internet and social media penetration, openness to democratic values and traditions as well as terms of technological innovations is a myth that has a degree of veracity and pertinence. Endong (2020) further indicated that in the digital age in which we are, through the explosion of the Internet and social media communications, blurred lines exist between truth and what should be considered false or faked and this in a way affected the smooth-running digital diplomacy and change in the global south countries. Wekesa (2020) is of the view that there is a lack of digital diplomacy policies and the absence of digital diplomacy research and publication efforts on the continent. Of utmost importance, Antwi-Boateng and Al Mazrouei (2021) stressed that the digital divide between the global north and the global south also has implications for digital diplomacy practices for development and social change.

THEORETICAL FRAMEWORK

This chapter is strengthened by the practice theory and explores *how* the digital diplomacy service could be better understood, and practiced in the fourth industrial revolution. This theory could be used to emphasise the use, impact, and practice of diplomatic services through digital media platforms by the global south countries. Thus, there is a need to provide a historical overview, and a gritty description of the theory as well as explain the selected principles that are pertinent to this chapter.

Practice Theory for the Digital Diplomacy Transformations

As an increasing quantity of diplomatic practices takes new and digitalised forms, various research and surveys on practice approach in international relations theory to the digital transformation of diplomacy is also swiftly escalating and intensifying in the global south societies. As, Ong'ong'a (2021) indicated that currently our lives have been extremely inundated with numerous digital media platforms that significantly impacted and transformed how usually people practice or engage in economic, social, governance, and political activities in their respective spheres and spaces. Furthermore, Hedling and Bremberg (2021, pp. 1) highlighted that the "global spread of the internet technologies has fundamentally reshaped societies in just a few decades. Whereas in international politics, digital media is forcing diplomats to rethink core issues of governance, order, and international hierarchy". Moreover, in this present era or revolution, societies talk about e-commerce, e-learning, e-governance, e-government, etc which is equally the use of electronic media platforms to practice business, trade, run the government, and public affairs.

Hedling and Bremberg (2021, pp. 1) further supposed that this "intersection of diplomacy and information technology has led to the emergence of new practices of digital diplomatic services" and interaction. These technological platforms are universal, free, and internationally recognised and tend to reach a broader masses of people in different spheres of life such as education, health, government, politics, and economy. In addition, research shows that due to the plethora of online and digital media platforms, there has been a serious disruption of diplomatic practices and processes as a whole. Ong'ong'a (2021) also highlighted that through the digital practice approach, many countries, states, and governments have set up departments or sectors that deal with issues of international interest. In addition, as a way to establish a relationship between nations, they also send individuals to represent their missions in foreign countries, these people normally act as a link between these nations to strengthen the relationship and ensure a free flow in communicating diplomatic mission statements (Adesina, 2017, Turianskyi & Wekesa, 2021). However, these changes in the process and practice have brought about massive challenges and opportunities in the use of digital media technology for diplomatic services, practices, and governance. As, Hedling and Bremberg, (2021, pp. 2) concluded that it is important for academics and researchers to conduct studies or research on "digital diplomacy that builds rapidly on practice theory that is used in the international relations" discipline. In the same vein, Ekengren (2018) digital practice approaches in international theory have for instance proved valuable for better understanding the dynamics of international security, social relations, and international public affairs. Hence, there is a need to establish how and to what extent diplomats from various nations can establish relationships through the use of *Facebook, Twitter*, and *Instagram* to comment, engage, debate, and negotiate on international affairs that affect their respective countries.

Furthermore, McCourt (2016) argues that the main purpose of practice theory in international relations is to detail the relationship that exists between continuity and change, agency and structure. The theory focuses on drawing a critical and rational logic in the global world of politics and the national domain (Adler, 2019). However, in response to that Hedling and Bremberg (2021) indicated that digital media technology has managed to bring in 'new actors' into the diplomatic practice field and discipline. This new way of practicing diplomatic service also changed how these actors do things, present and perceive themselves in the global space. Naylor (2020) further argued that digital diplomacy can be said to have disrupted traditional diplomacy because it is in many ways a self-ascribed experimental practice. Diplomatic actors are often aware that digitalisation involves taking risks and engaging with the unknown, which in turn is at odds with the perception that diplomacy should display foresight and be risk-averse (Shumba, 2020). The practice idiom can be used to describe a range of concrete phenomena from mundane aspects of local e-mail protocol to the ceremonial use of social media in state representation or increasingly structured activities of teleconferenced negotiations in international organisations (Eggeling & Nissen, 2021). Moreover, the rapid move to "Google meets diplomacy" "Microsoft team diplomacy" and "zoom diplomacy" in early 2020 as a result of the COVID-19 pandemic establishes how these practices can, at least temporarily, replace face-to-face diplomacy, alas not without its attendant difficulties. In a nutshell, the beckoning questions addressed in this chapter are; what is the importance of using social media platforms for diplomacy communication in the global south? what are the hurdles brought by the 4IR during digital diplomacy communication? Lastly, what are the probable strategies used to prevent digital diplomacy challenges by global south countries?

METHODOLOGICAL ASPECTS OF DIGITAL DIPLOMATIC PRACTICES

The methodology is the strategy that academic researchers use or apply to answer the research questions and address the specific objectives of the study or survey. Khan, Kunz, Kleijnen, and Antes (2003) argue that for any systematic review to become successful, the researchers should have the skills and capability to identify any successful systematic review to summarise their evidence by using clear practice. Bello et al., (2015) asserted that "a systematic review is defined as a review using a systematic method to summarise evidence on questions with a detailed and comprehensive plan of study". Consequently, this research followed a systematic review of the research published between 2012 and 2022 by searching various repositories such as Google Scholar, ResearchGate, ScienceDirect, Scopus, Embase, and PubMed. The research included scholarships that reported conclusions related to digital diplomatic services, the causes, use, and implications of the global communication revolution, diplomacy in the digital age, and those analysing the relationship between digital communication and diplomatic services in the global south. At the same time, research that reported or analysed the relationship between diplomatic service and traditional media (newspapers, television, radio, and magazines) in the global north was excluded. The research also followed the principle of the preferred reporting items for systematic reviews and meta-analysis (PRISMA) model in screening different publications, using appropriate keywords relevant to the topic. The Newcastle Ottawa Scale (NOS) was used to assess the quality of all cross-sectional studies included in this review. The research findings are discussed in the subsequent section.

DISCUSSION OF FINDINGS

The Main Benefits and Risks of Digital Diplomacy in the Fourth Industrial Revolution

The systematic analysis shows that in various global south countries such as South Africa, Nigeria, Kenya, China, South America, India, Mangolia, and Saudi Arabia; the fourth industrial revolution and digital media platforms which include social media, and new media technology present a path toward addressing diplomatic service and practice problems through the use of intelligent systems and the internet. The use of the internet and new media technology by global south countries came with various benefits which include but are not limited to improving their diplomatic services and practices; participation and involvement in international affairs; promoting inclusivity at such meetings and improving the interaction between policymakers and citizens of the continent; maintaining contact with audiences as they migrate online and harnessing new communications tools to listen to and target important audiences with key messages and to influence major online influencers.

Accordingly, it was further established that most of the global south countries have adopted digital technologies to enhance digital diplomatic practices and participation in local, national, and international virtual or online meetings, conferences, and dialogues. As Carvalho (2020) indicated that digital diplomatic practice is essential to promote inclusivity, for instance in the coordination of continental conferences, meetings, and forums, thereby improving interactions between policymakers and citizens on the continent and foreign affairs issues. Furthermore, it was discovered that in most global south countries, participation in negotiating the international governance of digital technologies is critical and crucial in mitigating a peripheral role in the international knowledge structure, ensuring transformational rather than transactional relations when it comes to the fourth industrial revolution. In addition, participation in these virtual or online meetings and conferences helps the country, diplomats, and government officials to keep abreast with what is happening in the global space and ultimately achieve their diplomatic goals and objectives for sustainable development. Moreover, Jaiswal et al., (2021) indicated that diplomacy has been an indispensable part of global affairs, as it is what makes relations between global nations. One easier way to connect with the world is digitally in this age of technology. The digital revolution has impacted all aspects of life including International relations.

Furthermore, Lowy Interpreter (2015) asserted that one of the essential benefits of digital diplomacy could assist global south countries to equally participate in international dialogues to advance their foreign policy goals, extend their international reach, and influence people who are not residents of their countries. The chapter further discovered that with the emergence of the internet, new media technologies, and other digital innovations such as social media, the conduct of the global south government diplomatic practices and affairs has been drastically transformed and improved. While these communications improvements have pushed governments and diplomats to respond rapidly to world issues and current health crises such as the outbreak of the coronavirus pandemic and recently Monkeypox. As indicated before, digital diplomacy entails using innumerable digital and new technology to access instantaneous information, being interactive online communication to keep track of the country's affairs and activities. As, Natarajan (2014, pp. 91) indicated that "in the age of mass communication technologies and new media, the public diplomacy initiatives utilised to communicate these narratives have gone digital to communicate foreign policy goals and decisions, construct a strategic narrative of Indian foreign policy and counter-narratives inimical to Indian interests." As Adesina (2017) concluded that digital media provide a platform or space for easy access, interaction, and engagement between the government and its citizens and thus, advancing the diplomatic goals and objectives.

Effective Use of Digital Media Technology for Diplomatic Practices During Covid-19

The use of digital diplomacy also allows traditional diplomatic practices and services to be delivered quicker and more cost-effectively. As Adesina (2017) highlighted that the use of digital media for diplomatic services and practices helps the global south countries conduct their businesses at ease and reduces traveling, office buildings, food, and accommodation costs and budgets. Furthermore, it is recorded that digital diplomacy is increasingly a part of relations between countries across the global south and beyond, including both developed and emerging economies of the world. Some of these relations remain transactional, where engagement in digital technology has been confined to achieving limited outcomes that have been short-term in nature. Of greatest importance, the power structure in the global south is reflected in the digital gap between the developed and developing countries, deepening questions of inclusion in the fourth industrial revolution landscape. Thus, there remains a shortfall in understanding the influence of digital technologies from the perspective of international relations and diplomacy, particularly on questions of equality, governance, and emerging transnational relations.

The deliberations in this section show that digital diplomacy has become an extremely sophisticated phenomenon with massive methodologies, technologies, and techniques at its disposal. However, like any other scope of the research area, digital diplomacy is not insusceptible or immune from criticisms and the impact that comes with using it to attend government meetings and conferences during the COVID-19 pandemic. In addition, diplomats, heads of state, foreign affairs people, and government officials are continually faced or confronted with various problems and challenges due to the ever-changing digital environment. Firstly, these digital media platforms are not well-regulated and controlled to advance and engage in diplomatic practices and services. Today, the internet space hosts more than 3 billion people, most of whom only access the internet through their mobile phones (Adesina, 2017). The majority of people can access digital space and create their accounts. Secondly, account hacking has become another problematic issue in the digital space. Anyone with an internet connection and bad intentions could easily hack into the government and diplomats' accounts and just post any information that could put the country or an individual into disrepute. Thirdly, in this dispensation, anonymous internet users have also emerged as one of the potential risks of digital diplomacy as any person using false details could just create an account on behalf of the country or foreign affairs department and equally, post false or inaccurate information about them. Fourthly, Misformation, and disinformation in digital media have come under the spotlight as false and inaccurate information can be disseminated using the name of the country or diplomat of a certain country. Since during this fourth industrial revolution, information is stored in the could, there is a possibility of important information leakage, cybercrime, and cyberbullying activities also happen.

Strategic Solutions to Prevent Digital Diplomacy Challenges in the Global South

This study established that the digital diplomatic challenges faced by global south countries and governments could be perfectly dealt with in totality. Endong (2020) highlighted that to come up with various solutions to digital diplomatic challenges faced by global south communities; diplomats and diplomatic missions should adjust to their "analogue" habits to be at the breast of the ongoing internet-driven transformations in the world and to be relevant in the new information sphere. In addition, Endong emphasised that the post-truth era has for instance motivated diplomats to get ahead of events and to respond to alternative truths/facts in real-time. Such an era should also force or warrants MFAs, embassies and foreign policy professionals to react to events particularly fake news on the fly to limit damages and protect the image and diplomatic standpoint of their countries or institutions on the national or international scene. In addition, Losifidis and Wheeler (2016) suggested that it is paramount important for the international arena to be reviewed to uncover the extent to which social media have constructed (or not) new public spheres within the global south. These would help in addressing the various challenges faced by diplomats and global south governments. As, Hocking and Melissen (2015) specified that the rise of networking sites like Twitter, Facebook, and other social media is important, but the ongoing debate equally needs to address the wider impact and challenges of digitalisation on the external relations of international actors such as diplomats, especially in the global south arena.

Furthermore, to curb digital diplomacy challenges in the global south, Hocking and Melissen (2015) proposed that departments of international relations and foreign ministries should define what they mean by 'digital diplomacy' in their context. The vivid argument is that digital diplomacy is a contextual-based phenomenon and should be clarified or defined accordingly in different global south countries. In addition, the term requires a greater degree of precision than is commonly given in government circles, which reminds of vague references to 'soft power' by political leaders and diplomats. In the same vein, diplomats need to apply their skills to disentangling and interpreting key arguments about digitalisation. Melissen and de Keulenaar (2017) are of the view that there is an urgent need to analyse digital technologies as mediating political processes, and thus of digital diplomacy as having its own "digitally native" forms. These might assist to deal with various challenges faced by global south countries with the use of digital media platforms for diplomatic services and practices.

Of utmost importance, in their studies, Hocking and Melissen (2015) emphasised that strategic solutions to prevent digital diplomacy challenges in the global south include but are not limited to evaluating key needs and resources in the digital field; promoting supportive internal structures such as digital units; establishing a 'mainstreaming' strategy whereby digitalisation percolates throughout the organisation; identifying and/or recruiting 'digital champions'; determining the key skills needed and modes of training to promote them; developing rules for using digital tools and guidelines on risk management; among others. Whereas, Masters (2021) indicated that global south states need to press for a transformational approach in their digital diplomacy, building relations as they negotiate new terms to overcome the digital divide and gain a place in the 4IR. In a nutshell, Melissen and de Keulenaar (2017) accentuate that diplomats should realise that digital diplomacy constitutes an engagement with how culture, information, and relations are systematised in software, such as with the counteracting of procedures that do not work in one's favour.

Implications of Digital Media for Diplomatic Practices in the Fourth Industrial Revolution

Given the centrality of communication in diplomacy, it is hardly surprising that the rise of social media should be of interest to diplomatic practitioners and government officials (Hocking & Melissen, 2015). Digital implication and adoption as stated by Bjola and Manor (2022) is an internally reflective process by which diplomats and diplomatic institutions try out and assess digital technologies and choose which ones to embrace in support of their foreign policy goals. Similarly to this, Hocking and Melissen (2015) indicated that an excessive focus on social media by diplomats conflates new communications technologies with broader dimensions of change in domestic and international policies. The shortcoming posed by digital technologies in diplomacy demand strategies dealing with the integration of 'online' and 'offline' environments (Hocking & Melissen, 2015). In this regard, the above-mentioned authors emphasised that the conventional wisdom among diplomats is that digitalisation does not change the fundamental objectives of diplomacy but offers new ways through which these can be achieved. As stated by Wekesa (2020) digital diplomacy is increasingly taking centre stage as a game-changing concept and practice of global affairs as we enter the third decade of the 21st century.

With the quick emergence of the fourth industrial revolution and the massive use of the internet between 2020-2022 due to the lockdowns and the coronavirus pandemic, most people might have perceived digital diplomacy as a new concept and practice in academia. However, as indicated above, digital diplomacy is a phenomenon that emerged in different sectors in the early 1980s and became the centre of desirability in the 21st century, especially in the year 2021 during the coronavirus lockdowns. In addition, the study conducted by Tham (2020) shows that digital diplomacy or commonly known as e-diplomacy started to gain momentum around 2001 through the global north or developed countries such as the United States of America, the United Kingdom, and some Commonwealth countries. It can be argued that digital diplomacy arrived on the African continent as part of the introduction of the internet in countries including Kenya, Nigeria, and South Africa in the 1990s. For instance, in response to the Covid-19 pandemic and associated lockdown in international travel, Nigeria's commission for diaspora affairs organised virtual meetings, provided online advice on evacuations, and advocated for Nigerian victims of human trafficking (Adesina, 2017). These expansions show that since its emergence, digital diplomacy had continued to slowly become part of daily activities and practices by diplomatic entities and diplomats to conduct foreign and diplomatic affairs. Hence it was generally important to focus on or consider subtleties within the practice of digital diplomacy in the global south countries, including the South African context.

The Future of Digital Diplomatic Practices in the Global South

Existing studies and literature show that through the emergence of the fourth industrial revolution, digitalisation has swiftly become part of our daily interaction. This discourse shows that the integration of digital media in rendering diplomatic practices or services has not transformed the way these services are offered by diplomats; but has enhanced their interconnectivity, and participation in diplomatic meetings, conferences, and decision-making podiums to assist their countries to achieve their nationally pre-determined SDGs. Additionally, the current diplomats also face the challenge of properly targeting audiences. It is imperative that in the future, diplomats understand whom they are targeting, what messages to send, and when to promote a certain message (Antwi-Boateng & Mazrouei, 2021). Furthermore, it is through these digital media platforms and the outbreak of the COVID-19 Pandemic that diplomats and government officials have been obliged to react swiftly to world events and take hasty decisions. As Ittefaq (2019) is of the view that social media and new technology open up new avenues for governments, individuals, and organisations to engage with foreign audiences and communities.

For instance, during COVID-19, global countries including the global south called for urgent virtual meetings to address the pandemic. We have witnessed the massive use of digital media platforms by government officials and diplomats to keep the public informed. Decisions were also taken via online or virtual platforms. With the exponential growth of media use, digital media have been of greatest assistance for diplomats to render their respective services, practices, and activities at ease. During these challenging times, it has become vividly evident that the global south countries have jointly embraced the use of digital media for diplomatic services in the era of digitalisation, information society, and the fourth industrial revolution. However, some authors such as Masters (2021) found that it is clear that digital diplomacy practised according to a transactional approach will leave global states excluded from the knowledge economy and the 4IR. Further, Ittefaq (2019) highlighted that digital diplomacy literature suggests that social media can help countries to build their positive image through engagement and dialogue. Furthermore, the above-mentioned authors suggested that countries must avoid across-the-board approaches and instead conduct market research to identify what media tools are best for targeting each country, what messages to deliver, and which audience to target.

CONCLUSION

Digital diplomacy remains the perceived tool or means of engagement between parties but as a means of navigating the evolving international digital governance regime and negotiating a more level playing field to address the inequalities in the fourth industrial revolution. Whereas, the outbreak of the COVID-19 pandemic had an intensive influence on the operation and practice of diplomatic services. In addition, intensifying the analysis of digital diplomacy thus helps shift attention to questions concerning the negotiation of access, resources, skills, and priorities of global south stakeholders in the fourth industrial revolution. This measures the current understanding of the concept of digital diplomacy before considering the context of the previous industrial revolutions in shaping the global south position. This paves the way for discussions on the role of digital diplomacy in the global south context, where transactional engagement between the countries and the international milieu reflects in the fourth industrial revolution. Yet, as this discourse points out, there is scope for the global south's participation in advancing a position on digital technologies that would challenge the status quo. This includes contributions to multilateral and regional negotiations that are shaping transformational relations around digital technology for the fourth industrial revolution. Moreover, it is recommended that there is a need to mediate the social tensions that arise from the introduction of digital technologies within global south countries because it is pivotal and necessary to address the impact of digital technologies on relations between countries.

REFERENCES

Adesina, O. S. (2017). Foreign policy in an era of digital diplomacy. *Cogent Social Sciences*, *3*(1), 1–13. doi:10.1080/23311886.2017.1297175

Adler, E. (2019). *World ordering: A social theory of cognitive evolution.* Cambridge University Press. doi:10.1017/9781108325615

African Union. (2020, September 10*). The digital transformation strategy for Africa (2020-2030).* AU. https://au.int/sites/ default/files/documents/38507-doc-DTS-english.pdf

Antwi-Boateng, O., & Al Mazrouei, K. A. M. (2021). The challenges of digital diplomacy in the era of globalization: The case of the United Arab Emirates. *International Journal of Communication*, *15*(1), 4577–4595.

Bello, A., Wiebe, N., Garg, A., & Tonelli, M. (2015). Evidence-based decision-making: Systematic reviews and meta-analysis. *Methods in Molecular Biology (Clifton, N.J.)*, *12*(1), 397–416. doi:10.1007/978-1-4939-2428-8_24 PMID:25694324

Bjola, C., & Manor, I. (2022). The rise of hybrid diplomacy: From digital adaptation to digital adoption. *International Affairs*, *98*(2), 471–491. doi:10.1093/ia/iiac005

Carvalho, G. (2020, June 11). *Africa must unmute its mic as e-diplomacy takes root.* ISS Africa. https:// issafrica.org/iss-today/africa-must-unmute-its-mic-as-e-diplomacy-takes-root

Eggeling, K. A., & Nissen, R. (2021). The synthetic situation in diplomacy: Scopic media and the digital mediation of estrangement. *Global Studies Quarterly*, *1*(2), 102–119. doi:10.1093/isagsq/ksab005

Endong, F. P. C. (2020). Digitization of African public diplomacy: Issues, challenges, and opportunities. *International Journal of Digital Society*, *11*(2), 1607–1618. doi:10.20533/ijds.2040.2570.2020.0201

Faye, M. (2000). *Developing national information and communication infrastructure policies and plans in Africa.* Paper presented at the Nigeria NICI workshop, Abuja, Nigeria.

Hedling, E., & Bremberg, N. (2021). Practice approaches to the digital transformations of diplomacy: Toward a new research agenda. *International Studies Review*, *1*(1), 1–24. doi:10.1093/isr/viab027

Hocking, B., & Melissen, J. (2015, July 12). Diplomacy in the digital age. *Clingendael Magazine.* https://www.clingendael.org/sites/default/files/pdfs/Digital_Diplomacy_in_the_Digital%20Age_Clingendael.pdf

Ittefaq, M. (2019). Digital diplomacy via social networks: A cross-national analysis of governmental usage of Facebook and Twitter for digital engagement. *Journal of Contemporary Eastern Asia, 18*(1), 49–69.

Khan, K. S., Kunz, R., Kleijnen, J., & Antes, G. (2003). Five steps to conducting a systematic review. *Journal of the Royal Society of Medicine, 96*(3), 118–121. doi:10.1177/014107680309600304 PMID:12612111

Kurbalija, J. (2017). The impact of the internet and ICT on contemporary diplomacy. In P. Kerr (Ed.), *Diplomacy in a globalising world* (pp. 151–169). Oxford University Press.

Losifidis, P., & Wheeler, M. (2016, May 13). *Public diplomacy 2.0 and social media.* Research Gate. https://www.researchgate.net/publication/303097870_public_diplomacy_20_and_the_social_media

Lowy Interpreter. (2015, April 15). Does India do digital diplomacy? *Lowy Interpreter.* http://www.lowyinterpreter.org

Manor, I. (2019). Digital diplomacy in Africa: A research agenda. *Hague Journal of Diplomacy, 10*(4), 538–574.

Masters, L. (2021). Africa, the Fourth Industrial Revolution, and digital diplomacy: (Re)Negotiating the international knowledge structure. *South African Journal of International Affairs, 28*(3), 361–377. doi:10.1080/10220461.2021.1961605

McCourt, D. M. (2016). Practice theory and relationalism as the new constructivism. *International Studies Quarterly, 60*(3), 475–485. doi:10.1093/isqqw036

McGinnis, D. (2020, October 27). What is the fourth industrial revolution. *Sales Force.* https://www.salesforce.com/blog/what-is-the-fourth-industrial-revolution-4ir

Melissen, J. (2015). *The new public diplomacy: Soft power in international relations.* Palgrave Macmillan.

Melissen, J., & de Keulenaar, E. (2017). Critical digital diplomacy as a global challenge: The South Korean experience. *Global Policy*, *8*(3), 294–302. doi:10.1111/1758-5899.12425

Mhlanga, D., & Moloi, T. (2020). Covid-19 and the digital transformation of education: What are we learning on 4IR in South Africa? *Education Sciences*, *10*(2), 1–11. doi:10.3390/educsci10070180

Moll, I. (2022). The fourth industrial revolution: A new ideology. *TripleC*, *20*(1), 45–61. doi:10.31269/triplec.v20i1.1297

Natarajan, K. (2014). Digital public diplomacy and a strategic narrative for India. *Strategic Analysis*, *38*(1), 91–106. doi:10.1080/09700161.2014.863478

Naylor, T. (2020). All that's lost: The hollowing of summit diplomacy in a socially distanced world. *The Hague Journal of Diplomacy*, *15*(2), 583–598. doi:10.1163/1871191X-BJA10041

Ong'ong'a, D. O. (2021). Systematic literature review: Online digital platforms utilisation by the ministry of foreign affairs in adopting digital diplomacy. *International Journal of Arts. Sciences and Humanities*, *9*(1), 8–18.

Ritto, L. (2014, October 18). Diplomacy and its practice vs digital diplomacy. *Diplomat Magazine*. http://www.diplomatmagazine.nl.diplomacy-practice-vs-digital-diplomacy-2.

Schwab, K. (2016). The fourth industrial revolution. *Geneva: World Economic Forum*. Springer.

Shumba, E. (2020, Ocober 09). Twiplomacy in Africa: Possibilities and pitfalls for diplomats. *Africa portal*. https://www.africaportal.org/features/twiplomacy-africa-possibilities-and-pitfalls diplomats/

Tham, D. (2020, January 09). Taiwan's digital diplomacy gets a kickstart. *Taipei Times*. https://www.taipeitimes.com/News/feat/archives/2020/01/09/2003728948

Turianskyi, Y., & Wekesa, B. (2021). African digital diplomacy: Emergence, evolution, and the future. *South African Journal of International Affairs*, *28*(3), 341–359. doi:10.1080/10220461.2021.1954546

Wekesa, B. (2020, July 03). Pathways for theorising African digital diplomacy. *Africa Portal.* https://www.africaportal.org/features/pathways-theorising-african-digital-diplomacy/

Wekesa, B., Turianskyi, Y., & Ayodele, O. (2021). Introduction to the special issue: Digital diplomacy in Africa. *South African Journal of International Affairs*, *28*(3), 335–339. doi:10.1080/10220461.2021.1961606

Zsubrinzky, Z. (2020). Digital communication in diplomacy. Paper presented in the *12*th *International Conference of J. Selye University Pedagogical Sections,* (pp, 147-155). J. Selye University.

Chapter 2

Technological Determinism and African Public Diplomacy:
A Conceptual Perspective on African e-Diplomacy in the Post-COVID-19 Era

Floribert Patrick C. Endong
iD https://orcid.org/0000-0003-1893-3653
University of Dschang, Cameroon

ABSTRACT

Although a galaxy of research works have focused on diplomacy in the COVID-19 period, very little attention has particularly been given to the idea of comparing the state of African digital diplomacy in the pre and the post COVID-19 period. Also, the literature available suggests that most studies on digital diplomacy in the COVID-19 period are Eurocentric and the COVID-19 pandemic as an opportunity for the future growth of digital diplomacy in Africa is still virtually understudied. In view of filling the above-mentioned gap in knowledge, this chapter deploys secondary sources and observations to attain three principal objectives. First, the chapter explores the state of digital diplomacy before the outbreak of the COVID-19 pandemic. Second, it examines how the pandemic positively affected the conduct of digital diplomacy in Africa, putting to question the myth of digital divide between the West and Africa. Lastly, the chapter explores implications for the practice of public diplomacy in post-COVID Africa as well as for future research on African digital diplomacy.

DOI: 10.4018/978-1-7998-8394-4.ch002

INTRODUCTION

The outbreak of the COVID-19 Pandemic has engendered a huge wind of change which has swept across all human industries in the world. The pandemic has actually given birth to a complex and difficult international environment, thereby justifying the adoption by governments of very radical policies (such as the lock-down of cities, travel restrictions and quarantine systems) to enable both human survival and industrial production in their respective territories. The outbreak also enabled the explosion of various digital cultures (notably e-administration, e-government and virtual meeting) which themselves have profoundly revolutionised almost all facets of life on planet earth. In tandem with this, one may observe that the COVID-19 created conditions favourable for almost all human activities to migrate online thereby enabling an explosion of cultures such as smart-cities, e-banking, e-administration, e-learning, e-commerce, and online advocacy among others. The above-mentioned developments have motivated many scholars to describe the COVID-19 period and the post-COVID period as an era of the new or next normal (Peru, 2022).

Like all the above-mentioned human industries, public diplomacy has been seriously affected by the pandemic. This is evidenced by at least two things; the first is the fact that embassies intensified their online presence with trepidation, in order to provide consular services to stranded nationals, strengthen bi-and-multilateral cooperation, and maintain their diplomatic footprint. The second evidence is that most of the international government summits, conferences, and gatherings organised during the outbreak were held through teleconferencing and other digital platforms. The option of using such tele-conferencing and digital platforms for international gathering has since then remained popular and a normality. According to the UN headquarter in Geneva, the United Nation Organisation ran over 1200 important international conferences online between March and the end of the year 2020 (United Nation Organisation, 2021).

A plurality of scholars have sought to show how the pandemic redefined the conduct of diplomacy in the world (Abdelhafidh, 2021; Internet Governance Forum, 2021; France Diplomatie, 2022). According to Labott (2022), the pandemic caused the diplomatic crossroads of the world to go into "hibernation". With close reference to the American experience, Labott (2022) explains that the culture of flocking to Washington for meetings at the White House, State Department or World Bank suddenly gave way to diplomacy enabled by video-conferences and phone calls. In her words, "world leaders and ministers [were] forced to conduct diplomacy over video conferences and phone calls, while ambassadors [carried] the load of representation and advocacy. Embassies and diplomatic residences, once the hubs of envoys from the U.S. and around the world gathering for cultural events and receptions, are now eerily quiet" (p.2). In the same line of argument, Bjola and Manor (2022) observe that the COVID-19 pandemic forced world diplomats to work online and develop hybrid approaches to the conduct of public diplomacy. Hybridity here means the adoption of both virtual and physical meetings in their works. Most public diplomacy scholars also observe that the effects of the COVID-19 pandemic on human interactions and industries suggest that people-to-people interaction are bound to be lost and embassies in some years/decades ahead are also likely to look quite different from now (Puru, 2022; Labott, 2022).

The sources mentioned above suggest that a number of scholars have investigated the impact of the covid-19 pandemic on the conduct of public diplomacy in the world. However, in spite of the availability of the research works mentioned above, scholarly attention to the impact of the COVID-19 pandemic on digital diplomacy has mainly been on western countries leaving African experiences grossly understudied. Only few authors such as Wekessa (2021), Adesina (2022), Manor (2021) and Endong (2020) have addressed a handful of issues pertaining to the influence of the COVID-19 pandemic on digital diplomacy in Africa. Thus, the literature available suggests that the COVID-19 pandemic as an opportunity for the growth of digital diplomacy in Africa is still virtually understudied, and thus an interesting topic for scholarly discourse. In view of filling the above-mentioned gap, this chapter deploys a critical exploration of secondary sources and observations to attain three principal objectives. In the first place, the chapter examines the state of digital diplomacy before the outbreak of the COVID-19 pandemic. In the second place, it critically examines the state of African digital diplomacy in the pre-COVID-19 period. In the third place, the paper studies how the pandemic positively affected the conduct of digital diplomacy in Africa putting to question the myth of digital divide between the west and Africa. In the last place, the chapter explores implications for the practice of public diplomacy in post-COVID Africa. In line with the above objectives, the paper is divided into four main parts. The first part provides a theoretical framework which is composed of technological determinism and digital divide. The second part explores the state of African digital diplomacy in the pre-COVID period. The third part examines the impact of COVID-19 on the growth of African digital diplomacy while the last part addresses the future of digital diplomacy in Africa.

THEORETICAL FRAMEWORK

This paper is anchored in two theories namely technological determinism and the digital divide concept. These theories could be used broadly for a variety of information technologies-related studies. Thus, there is a need to provide a granular definition of the two theories as well as explain the selected tenets that are relevant to this chapter.

Technological Determinism

In its strict sense, technological determinism refers to the popular proposition that technology determines the state of human industries in a society. According to Webster (1995, p.39), it is a theory which present technologies as being aloof, yet decisive social agents. By this theory, technology is regarded as the driving force of a society to the point that it determines the course of this society in history. There are various versions of this theory. One such version stipulates that technological progress is bound to engender newer ways of production in a society. These newer ways in turn are likely to influence the cultural, political and economic aspects of a society thereby engineering a transformation of the society in question. Langdon Winner (2016) supports the above version of the technology determinism theory. He advances two hypotheses as follows: (a) technology is a fundamental influencer of the various aspects of life in a given society and (b) changes in technology are the principal catalysis of social change in a culture. In the specific domain of digital diplomacy, technological determinism has similarly been defined in various terms (Leach 2015; Bjola & Holmes 2015; Comor, 2017; Bjola, 2017; Decos, 2018). However, this chapter will hinge specifically on Hocking and Melissen's (2015) conceptualisation of technological determinism in the book titled "Diplomacy in the Digital Age".

The two scholars theorise on how new technological developments affect the conduct the diplomacy in the world. They advance a plurality of ideas/hypotheses related to diplomats' reception of technological innovations. Three of such hypotheses that will be of interest to this chapter are discussed as follows:

i.   MFAs and diplomats' reception of developments in communication (particularly digitalisation) as well as their appreciation of the importance of these technologies for the conduct of diplomacy mostly moves through three successive stages: (a) the phase where there is a mix of scepticism and hype, (b) the gradual acceptance of the technologies and (c) the mean-streaming of the technologies within organisations. According to most MFAs were, as at 2015 entering the digital age and were at the first stage.

ii.   Diplomats will always find the modalities of digitalisation in constant flux. This will compel them to retool on a continuous basis. Most of what is considered revolutionary in the practice of digital diplomacy will soon be regarded as being commonplace or outdated.

iii.   The gap between government that do not invest in understanding the impact of digitalisation on diplomacy and those that do will increase with the speed and velocity that are typical of the digital age. (Hocking & Melissen, 2015)

Hocking and Melissen's (2015) third point mentioned above is (somewhat) reminiscent of the digital divide between Africa and the West. By the above hypothesis, this digital divide is likely to widen in the digital age and determine the differences between African digital diplomacy and the western one. This digital divide is the second theory on which the chapter hinges.

The Digital Divide Theory

The theory seeks to describe two things: (i) disparities in accessibilities to communication technologies and (ii) the abilities to use these technologies. The OECD (2011) defines the term as the "gap between individuals, households, businesses and geographic areas at different socio-economic levels with regard both to their opportunities to access ICTs and to their use of the Internet for a wide variety of activities". Digital divide is in theory revealed by three indicators namely (a) information accessibility, (b) information utilization and (c) information receptiveness. On the bases of these three indicators, two types of digital dived have been identified. The first (called first order digital divide) has to do with access to information and communication technologies and cost associated with going online. This type of digital divide is determined by a variety of factors one of which is the varying rate of internet and social media penetration across countries of the world. According to many sources for instance, internet and social media penetration is lower in African countries, compared to other parts of the world. The uneven spread of the internet and social media has been used as an index to illustrate digital divide between rich and poor countries.

The second type of digital dived (called second order) is associated with digital literacy and digital capabilities. Here again, studies have revealed that second order digital literacy differentiate rich from poor countries (Alliance for Affordable Internet, 2016; Tara, 2022;). Digital literacy and capabilities are theoretically higher among western MFA staffers compared to African ones. Manor (2015) notes for instance that:

Younger diplomats in Europe or North America may have opened a Hotmail email account while they were in Middle School and may have even used PCs throughout most of their education. This is not necessarily the case with MFA staffers in smaller and poorer African nations. Given that length of familiarity with the online environment may substantially increase one's digital capabilities and literacy, digital diplomacy may still be characterized by second order digital divides. As a result, African MFAs may find it harder to fully harness the potential of social media and use it as an effective tool for information gathering and policy making. (p.4)

According to Hocking and Melissen (2015) the level at which countries invest in understanding the impact of digitalisation on diplomacy is bound to widen given the speed and velocity of change in the digital age. This suggests that as long as poor countries (notably African countries) will fail to invest in critical technology as well as in conceiving digital policies and strategies, they will continue to lag behind the West.

STATE OF AFRICAN DIGITAL DIPLOMACY IN THE PRE-COVID-19 PERIOD

Since the early part of the 2000s, there has been a growth in internet penetration as well as an increasing proliferation of the mobile telephony and smart phones in Africa. There has also been a remarkable emergence of a variety of internet related cultures (from e-government to online activism and e-learning) on the continent. This development has pushed many observers to argue that, although still lagging behind the West, Africa had since embraced the majority of digital cultures that had originated from the West (Internet Governance Forum, 2021; Endong 2018; 2017; Manor 2016; 2018; Adesina,

2022; Allen, 2022; Tara, 2022). In effect, there is no doubt that Africa before the COVID-19 pandemic was home to a majority of internet and social media driven cultures. In line with this, a good number of researchers claim the digital divide between Africa and the West has not really hindered African diplomats or MFAs from practicing some forms of digital diplomacy as from the early part of the years 2000s. In a 2015 study aimed at comparing a selected number of African MFAs and their Western counterparts in terms of use of ICTs in the conduct of public diplomacy, Kampf et al (2015) argue that African diplomats were, as of the time of the investigation, as active as their western counterparts. In some cases, African MFAs were even more active than their western counterparts. Kampf et al (2015) write that:

African MFAs [considered in the study] were found to be among the most active on SNS [Social Networking Sites], suggesting a narrowing of some aspects of the digital divide. While African countries may lag behind Western countries in terms of internet penetration, computer infrastructure and internet accessibility, African governments seem to be equally active and as committed to engagement as their Western counterparts, if not more so. (p.13-14)

Another study conducted by Manor a year later sought to demonstrate how African MFAs – notably those of Namibia, Kenya, Niger, and the Democratic Republic of Congo among others – have been using the social media to engage foreign audiences and popularise their activities and foreign policies. In his paper titled "Digital Diplomacy in Africa: A Research Agenda", Manor (2016) argues that such use of the ICTs in the conduct of public diplomacy has brought additional value to African countries in such areas as diaspora diplomacy, networked diplomacy and nation branding. Thus, pre-COVID 19 studies devoted to African digital diplomacy have established that in the area of digital diplomacy, the digital divide between the West and Africa may be narrowing considerably. The study also reveals that a number of African MFAs and diplomats have long ago adopted the culture of creating websites or social media handles and of being active to some extent on the internet.

A 2019 survey conducted by Twiplomacy confirms the above observation. This survey reveals that the practice of deploying Twitter for diplomatic communications had become popular among a good number of African leaders and policy makers, long before the the outbreak of the COVID-19 pandemic. The survey stipulates for instance that African politicians such as Rwanda's Paul Kamgame, Kenya's Uhuru and Ghana's Akufo Ado had, before the pandemic, been making visible efforts towards using the social media platforms such as Twitter, Facebook and Instagram for diplomatic purposes. The survey shows that while the Kenyan president has a very large number of followers on Twitter (7 million followers), Rwanda's Paul Kagame is amongst the most conversational Heads of States on Twitter. Kagame has 2.5 million followers on Twitter, Facebook, and Instagram combined. The same study reports that Ghana's Akufo-Ado has 421k followers on Instagram.

In spite of the above-mentioned positive development, the state of digital diplomacy in pre-COVID 19 Africa was somewhat deplorable. In other words, all has not been well with digital diplomacy on the continent, particularly in the period before the outbreak of the pandemic. Africa has continued to represent the lowest spot of digital diplomacy. This has been so for reasons which range from low internet penetration and low democratic culture to limited internet presence of diplomatic officials and reduced digital literacy among African diplomats. Although a number of African diplomats, MFAs and heads of States have been enjoying a degree of internet presence before the COVID-19 pandemic, very few had really been active on the internet or the social media. In its 2019 survey, Twiplomacy observed that contrary to Ghanaian, Kenyan and Rwanda presidents who had been digital diplomacy enthusiasts and who used to enjoy huge followings on social media, most African politicians and policy makers still lag behind in the use of the most basic forms of digital diplomacy. For instance, the leaders of Eritrea, Mauritania and Swaziland had no Facebook account. Similarly, the MFAs of many other African countries used the social media in a way that was rather mundane and not exclusively for foreign policy and diplomatic communications. In most cases, communications initially conceived for the consumption of domestic publics were used for diplomatic purposes when needs arose. It is unfortunate that, the trend continued to some extent to be perceived even after the outbreak of the COVID-19 pandemic.

In addition to the above, African countries have generally been lagging behind when it comes to formulating digital diplomacy policies or creating bodies that are formally saddened with the duty of (re)thinking their digital diplomacy policies. Contrary to Western countries that have since the early part of the 2000s put in place structures in charge of guiding their application of digital diplomacy, most African countries do not have clearly stated digital diplomacy policies or strategies. The very few exceptions have been South Africa, Kenya and Uganda (Wekessa, 2022; Adesina, 2022). In other words, the greatest majority of African countries have, particularly before the COVID-19 pandemic, not thought of putting in place African replicas of US's Digital Diplomacy Department (recently re-launched as US's Bureau of Cyberspace and Digital Diplomacy) or African replicas of the Swedish culture of designating some of its diplomatic officials as digital diplomats. In a nutshell, one may say that African countries have before the COVID-19 pandemic given the impression not to have fully appreciated the importance of digital diplomacy and seen the necessity to design formal strategies and policies to guide their practice of digital diplomacy. Before the outbreak of the pandemic, African diplomats tended to practice digital diplomacy instinctively rather than following well-conceived strategies and policies. As shall be discussed in subsequent parts of this essay, this absence of policies and strategies continues to be one of the greatest weaknesses of African MFAs and diplomats' use of social networking sites and the internet in their conduct of public diplomacy.

IMPACT OF THE COVID-19 PANDEMIC ON AFRICAN DIGITAL DIPLOMACY

The COVID-19 pandemic came with a series of disruptions that, for the most part, negatively affected most human industries in the world including Africa. The pandemic has been a threat to most economies on the planet as it has given birth to a situation of public health and economic precariousness which has in turn motivated the adoption of drastic measures such as lockdowns, travel restrictions and systems of quarantine. These measures have affected almost all human industries in Africa in a mainly negative way. The conduct of diplomacy in Africa – like in other parts of the world – has seriously been affected. However, contrary to other human industries, African digital diplomacy has positively been affected by the pandemic. Indeed, the COVID-19 pandemic has compelled African countries, diplomats, and MFAs to more and more apply the ICTs in their conduct of diplomacy, so much so that most African scholars have described the advent of the COVID-19 pandemic as a blessing and opportunity for African digital diplomacy. Wekessa (2022) notes for instance that the paradoxically positive effects of the COVID-19 pandemic on African digital diplomacy give credence to the African maxim that states that "every misfortune is a blessing," the African replica of the British dictum which stipulates that "every cloud has a silver lining."

Actually, the COVID-19 pandemic compelled the majority of African governments and diplomats to catch-up; the latter saw the necessity to emulate, with trepidation their western counterparts in the practice of digital diplomacy. The austerity, restrictions and regime of social distancing engendered by the pandemic seem to have forced virtually all African countries' MFAs, embassies, and diplomats to undergo the great trek of digital diplomacy. Even countries which had most often been lukewarm or phobic to the use of sophisticated forms of digital diplomacy seem not to have had any better choice but to make digital diplomacy one of their premier public diplomacy "fetishes". As rightly observed by Wekessa (2022) African MFA and diplomats' resort to ICTs for the conduct of public diplomacy has not been by choice, but by subtle obligation. Wkessa (2022) explains that:

With the arrival of the pandemic, African actors in fields of foreign affairs and international relations were forced to enhance their appropriation of digital technologies as they had few if any options for conducting core diplomatic functions. If African diplomatic actors were apprehensive, hesitant, or capacity-challenged in using digital technologies, they were constrained to embrace these technologies with alacrity. In what Ilan Manor characterised as the digitalisation of public diplomacy, African ministries of foreign affairs essentially leapfrogged from comparatively low to heightened levels of the use of digital technologies. (p.105)

In tandem with this, African presidents, diplomats and MFAs have with trepidation, learned to integrate digital tools to their *modus operandis*. Since the outbreak of the pandemic in 2020, most of them (African diplomats and MFAs) have for instance been forced to design their bilateral or multilateral meetings in virtual mode. Thus, concepts such as virtual conferences, virtual meetings and tele-conferencing have "overnight", and due to the unpredictable constraints of the COVID-19 pandemic, become more standard ways of designing negotiation and communication fora. A prime illustration of this prompt resort to sophisticated forms of digitally assisted negotiation and communication fora is the Extraordinary Session of the ECOWAS Heads of State and Government which held on the 22 to 23 of April 2020. This conference was held through teleconferencing (Economic Community of West Africa [ECOWAS], 2020). Another good example is the Heads of States Consultative Meeting of the East African Community which was similarly held through teleconferencing on May 12, 2020 (East Africa Community, 2020). Other examples include South African president Cyril Ramaphosa who attended the March 2020 edition of the G20 Summit not physically but virtually. Still in the early stage of the pandemic, the African Union was forced to leverage digital platforms for the running of its summits. A case in point is its March 2020 summits chaired by Cyril Ramaphosa the then chairman of the Union, and aimed at presenting African heads of States and government's response to the pandemic. A similar online-based meeting that regrouped African heads of State and government is the virtual heads of State summit the then newly elected US president Joe Biden held with African presidents and prime ministers in February 2021.

The resort to digital communications has had a very positive impact for the efficiency and transparency of diplomatic activities in Africa. In effect, African diplomats' use of social networks such as Instangram, Facebook and Twitter for the publication of videoconferences as well as for interactions with foreign publics made diplomatic activities to be more democratic. Thus an area that has long escaped the control of public opinion was now accessible to the common man in many African countries. In Nigeria for instance, politicians and senior diplomats who sought to address the maltreatment of Nigerians in China, entrenched the culture of publishing their meetings with Chinese diplomats on social media. These enabled the Nigerian populace as well as some foreign entities to be at breast of Nigerian government's reaction to China's xenophobic treatment of African Diasporas in its territory.

Besides the organisation of virtual summits, African MFAs and diplomats deployed the social media to track their citizens stranded in specific parts of the world and provide the latter with prompt and valuable consular assistance/ service. Many African Ministries of Foreign Affairs used these internet-driven platforms particularly to manage their citizens' repatriation. A good example is South Africa's Ministry of Foreign Affairs which, in collaboration with the Chinese government and the South African Defence Forces leveraged the internet to track South Africans living in Wuhan (China) and enable their safe repatriation in South Africa. Other countries such as Nigeria used similar approaches to repatriate their citizens.

Thus, the COVID-19 pandemic somehow forced African countries' leaders, diplomats, embassies, and MFAs to leap into an era of digital diplomacy, embracing en masse and with trepidation, Internet-based traditions (notably virtual meetings and virtual multilateral conferences) that had long ago become common in western countries. The pandemic thus enabled African countries to quickly effect a diagnostic of their diplomacy and to somehow see the need for greater efforts towards a more consequential embrace of digital diplomacy. The simple fact that most African critics and even diplomats described African MFAs and embassies' prompt resort to paradigms such as virtual multilateral and bilateral meeting and virtual conferences as a kind of novelty or as an unprecedented change should tell one that African diplomacy has for years been stuck into the old paradigm-based use of conventional channels.

THE FUTURE OF DIGITAL DIPLOMACY IN AFRICA

Discussion and theorisation on the future of digital diplomacy in Africa can be summed up in two terms namely (i) the imminence of hybrid diplomacy and (ii) the high probability that Africa will continue to lag behind the West in terms of the application of digital diplomacy.

The Imminence of Hybrid Diplomacy

The outbreak of the COVID-19 pandemic has made most African countries appreciate the necessity – nay imperative – of promptly embracing digital diplomacy among other digital cultures. The difficult national and international environment birthed by the pandemic warranted the use of internet-based innovations and virtual engagements for the safe conduct of public diplomacy. However, the relaxing of mobility and physical gathering restrictions as well as the possibility of organising face-to-face diplomatic engagements are today present. This development has pushed both commentators and scholars to wonder what the future of digital diplomacy in the world will be, as well as in Africa. It goes without saying that the COVID-19 pandemic has catalysed the emergence of what Bjola and Manor (2022) call a "*hybrid diplomacy*". This diplomacy is a context in which physical and virtual engagements integrate, complement, and empower each other. This type of diplomacy is bound to be present even in Africa in the post-COVID period.

Indeed, from many indications, digital diplomacy has come (to Africa as well as to the world) to stay and evolved into a more complex phenomenon – a *hybrid diplomacy*. At least two factors suggest that there will be a strong symbiosis between digital and analogue diplomacy in the years to come and in the post-COVID 19 era. The first has to do with the fact that digital diplomacy is in no way the "infant" of the COVID-19 period. This suggests that the longevity and legitimacy of digital diplomacy do not really depend on the COVID pandemic or any similar disruptive phenomenon. In effect, like all the other forms of digital cultures, e-diplomacy saw the light of the day many years before the outbreak of the COVID-19 pandemic. This digital culture was popular in Western countries as early as the first part of the year 2000s, as seen in the various Western countries' adoption of digital diplomacy policies and strategies. Sweden launched its "Virtual Embassy to the Second Life" as far back as 2006, while the US developed its "Digital Outreach Team in 2007 (Manor, 2018). African diplomats and MFAs also practiced some forms of digital diplomacy well before the outbreak of the Corona virus.

Before the pandemic, African countries such as Rwanda, Kenya and Uganda had already conceived policies for digital disporas outreach (Manor, 2018).

The COVID-19 pandemic simply brought to the fore the centrality of digital diplomacy and some other digital cultures. In an era of difficult mobility, improbable physical gathering, difficult working conditions and limited budgets, digital diplomacy offer diplomats a way forward. The difficult working conditions made the growth of digital diplomacy in Africa just inevitable. In a way, the COVID-19 pandemic justified the search for alternative ways of conducting public diplomacy in Africa. As noted by the United Nations Organisation (2021), "Diplomats did not discover on-line conferences with the pandemic. What has changed is the frequency of use of these tools" (p.16). Thus, there has been a kind of hybrid diplomacy since before the coming of the COVID-19 pandemic. This hybrid diplomacy is bound to subsist and become more complex in the post-COVID 19 era in Africa.

In line with the above, it could be argued that the pandemic simply motivated African countries that lagged behind in terms of application of digital diplomacy to massively and promptly embrace modern approaches and digital tools in their conduct of public diplomacy. In view of this, one may rightly contend that the post-COVID period calls for a synergy between physical and virtual diplomacy, not a decline of one form of diplomacy to the profit of the other. Thus, virtual diplomacy and analogue approaches to conducting public diplomacy are bound to integrate each other in post-COVID Africa.

The second factor that pleads in favour of a *hybrid diplomacy* in Africa is the fact that digital diplomacy may not be the best approach to handling a number of thorny diplomatic issues that compel active negotiations. In effect zoom diplomacy may prove inadequate for issues such as conflict resolution and peace talks among others. Such issues may better be addressed through face-to-face engagements. As noted by Secretary General of the Norwegian Refugee Council, Jan Igeland, direct discussions are the best approach to bringing together warring factions. As he puts it, "When the belligerent parties or their representatives meet each other, it is a gesture of great symbolic value, and which proves good faith" (cited in United Nation Organisation 2021, p.17). In the same line of thought, the director at the Forests, Land and Housing Division of the UN Economic Council, Paola contends that "making decisions is another story, and when the intergovernmental processes require negotiations and exchanges between delegates, online meetings don't make the diplomats' task any easier". In view of all these, one may argue that digital and analogue forms of diplomacy are bound to complement and integrate each other, especially in situations of conflict resolution. Actually issues such as peace talks are best resolved by people sitting round the table together. Adesina (2022) makes allusion to this necessary and predictable symbiosis between virtual and physical engagements when she writes that the successes of digital diplomacy imply in no way that it should supplant traditional face-to-face interactions. Adesina writes:

Instead, they [virtual diplomacy and face-to-face engagements] should be complementary. Physical meetings are important because negotiations often take place on the sidelines of international gatherings, through impromptu discussions between global leaders at tea or lunch breaks, or a chance encounter in the corridor or rest room and so on. Also, physical meetings provide an opportunity for participants to observe and interpret body language and emotions of the parties, which may help in decision making. Thus, hybrid format of physical interactions and online meetings seem to be the best approach for diplomatic engagements. Meetings should be held physically as they become more focused on decision making and high-level representation, such as issues that demand high level of secrecy, involve conflict situations, or complex negotiations. (p.95)

The application of ICTs in almost all human industries has become the norm. However, physical interactions still have their place in these human industries, diplomacy making no exception. This scenario strongly pleads in favour of hybrid diplomacy.

Africa Might Continue to Lag Behind the West

In their theories around the future of digital diplomacy in the world, Bjola and Manor (2022) contend that adhesion to the concept of hybrid diplomacy will among other things depend on countries' level of technological development and democratic culture. In their language:

The pace and shape of diplomacy will depend on how well MFAs manage the transition from adaptation to adoption; that is, from learning how to integrate physical and virtual presences under pressure, by trial and error and improvisation, to doing so in a more deliberative, strategic and systematic manner. For some, hybridity will probably remain a desirable aspiration hindered by technical challenges and institutional resistance. For others, hybrid diplomacy may well become second nature, allowing them to pursue their foreign policy goals fast, effectively and with confidence. (p.472)

Many factors suggest that most African countries will, at least for a long time, be in the first category mentioned above; that is the category of nations who regard hybrid diplomacy as a desirable aspiration hindered by technical challenges. This category is visibly a disadvantaged one. The factors that strongly suggest this future state of African digital diplomacy include issues such as the persistent low internet penetration in the continent, the persisting digital divide, the low African investment in ICTs and more especially the absence of digital policies and strategies as well as the absence of efficient training programs for African diplomats. According to scholars such as Tara (2022) and Internet Governance Forum (2021), Africa is the less digitised continent. It is also the continent in greatest need to develop its critical infrastructures. In the same line of thought, Wekessa (2022) lowly rates African governments in terms of the adoption of digital diplomacy policies and strategies. He contends that: "the point of departure is that digital diplomatic practices in Africa are not organised, deliberate, and purposeful. Rather, they are practiced inadvertently and haphazardly. This claim can be ascertained by looking at the digital diplomacy platforms of African ministries of foreign affairs, from Facebook pages to websites – which are often witness lows and highs in terms of the posting and uploading of content" (p.108).

Although Manor (2018) has observed that countries such as Rwanda, Kenya and Uganda have conceived policies for digital diaspora diplomacy; the dominant tendency among African MFAs has been the application of some forms of digital diplomacy without a formal and well defined policy or strategies.

The persistence of the above mentioned problems (absence of policies and the prevalence of some technical challenges) put Africa at a disadvantaged position and pleads in favour of the belief that Africa will continue to lag behind the west for a long time. In other words, while digitalisation of diplomacy or hybrid diplomacy will be the norm (particularly in the west), analogue diplomacy will continue to be dominant in Africa, unless African government surmount the multiple challenges that hinder virtual diplomacy on the continent. Of course, the above-mentioned problems and challenges call for greater efforts on the part of African governments. The latter actually need to draw inspiration from the West, in terms of investment in ICTs and development of digital policies and strategies. There is also the need for

African government to invest in the training of their diplomats, as well as for technological development.

CONCLUSION

From many indications, the future of African digital diplomacy is not as bleak as an Afro-pessimist observer may want to think, though a number of issues suggest that African countries are visibly at a disadvantaged position. The low levels of Internet and social media penetration, the poor digital diplomacy policies and the prevailing second level digital divide between African diplomats and their counterparts from the west strongly suggest that virtual diplomacy in the continent may not be as advanced as it is in western climes. However there exists a handful of concrete proofs which also suggest that African governments, MFAs and diplomatic missions are increasingly seeing the need to embrace more sophisticated forms of digital diplomacy. The current COVID-19 pandemic has obliged many of these African governments, MFAs and diplomatic missions to rethink their attitude towards digital diplomacy. The pandemic has in effect, forced them to leapfrog into an era of digital diplomacy where they know they must go beyond the usual child step of just creating websites and Facebook or Twitter accounts that are poorly designed, not regularly updated and not regularly used to engage with foreign publics and Diasporic communities.

The COVID-19 pandemic coupled with the acceleration of the smart society concept has plunged African countries into the pool of digital diplomacy, and overnight, the world saw how African heads of government, senior officials and diplomats scrambled to make virtual bi and multilateral meetings as well as teleconferencing common idioms and conventional tools in their practice of diplomacy. The adoption of lockdowns and social distancing as a result of the outbreak of the COVID-19 pandemic forced African countries' embassies to seriously think of rooting the bulk of their routine in the use of ICTs. Thus, in such a short span of time, many African countries embraced WhatsApp-based approaches as well as other e-tools to attend to their citizens' needs.

Although the COVID-19 pandemic spurred African government towards seeing the imperative to fully adjust and towards being at breast of the current digital revolution, all is still not well with African digital diplomacy. The industry faces numerous challenges such as limited internet penetration, poor cyber laws, inexistent digital diplomacy policies, digital illiteracy among diplomats and a relatively poor democratic culture. All these challenges will need to be addressed to enable a greater digitalisation of public diplomacy in the continent in the future. The persistence of the above-mentioned problems plead in favour of the negative belief around the future of digital diplomacy in Africa. One of such belief stipulates that while digitalisation of diplomacy will be the norm in the west, African countries will continue to dominantly apply analogue diplomacy. Issues such as the absence of clearly defined digital policies and strategies, the persistence of authoritarian cultures and the inexistence of reliable and adequate training for diplomats in the continent suggest that African countries will for a long time continue to lag behind the west in terms of application of digital diplomacy.

REFERENCES

Abdelhafidh, A. (2021). Digital diplomacy in an era of COVID-19. *Swiss Info*. https://www.swissinfo.ch/eng/business/digital-diplomacy-in-the-era-of-covid-19/46374914

Adesina, O. (2022). *Africa and the future of digital diplomacy. Foresight Africa: Technological innovations: Creating and harnessing tools for improved livelihood.*

Allen, K. (2022). *Cyber diplomacy and Africa's digital development (Africa Reports 38)*. Institute for Security Studies.

Alliance for Affordable Internet. (2016). World off-track to achieve UN internet access as broadband prices remain high. New York: Alliance for Affordable Internet (A4AI).

Bjola, C. (2017). *Trends and counter-trends in digital diplomacy. Working paper no18. "Digital Diplomacy in the 21st Century" project*. German Ministry of Foreign Affairs.

Bjola, C., & Holmes, M. (2015). *Digital diplomacy: Theory and practice*. Routledge. doi:10.4324/9781315730844

Bjola, C., & Manor, I. (2022). The rise of hybrid diplomacy: From digital adaptation to digital adoption. *International Affairs*, 98(2), 471–491. doi:10.1093/ia/iiac005

Comor, E. (2017). Technological fetishism and US foreign policy: The mediation role of digital ICTs. *The Political Economy and Communication*, *5*(2), 3–21.

Decos, A. S. (2018). Digital diplomacy and social capital: Analysing relational components of trust in US & Israeli online social media. [PhD Thesis, University of Otago].

Diplomacy, F. (2022). COVID 19: Assistance for Africa. *France Diplomatie*. https://www.diplomatie.gouv.fr/en/country-files/africa/news/article/covid-19-assistance-for-africa

Endong, F. P. C. (2020). Digitisation of African public diplomacy: Issues, challenges and opportunities. *IJDS: International Journal of the Digital Society*, *11*(2), 1607–1618. doi:10.20533/ijds.2040.2570.2020.0201

Hocking, B., & Melissen, J. (2015). *Diplomacy in the digital age*. Clinhendeal Institute.

Internet Governance Forum. (2021). *IGF 2021-day 2 – WS# 18, Cyber diplomacy in Africa and digital transformation*. IGF. https://www.intgovforum.org/en/content/igf-2021-ws-18-cyber-diplomacy-in-africa

Labott, E. (2022). *Redefining diplomacy in the wake of the COVID-19 pandemic*. The Meridian Center for Diplomatic Engagement.

Leach, J. (2015). Diplomacy – facing a future without borders. In Digital diplomacy & the # G8, (pp.15-17). London: Portland.

Manor, I. (2015). Between two digital divides, Exploring digital diplomacy. *Digital Diplomacy Blog*. https://digdipblog.com/2016/01/27/between-two-digital-divides/

Manor, I. (2018). *The Digitalization of diplomacy: Toward clarification of a fractured terminology Working Paper no 2*. Oxford Digital Diplomacy Research Group. https://www.qeh.ox.ac.uk/sites/www.odid.ox.ac.uk/files/DigDiploROxWP2.pdf

Organization for Economic Co-operation and Development. (2011). Understanding the digital divide. In *OECD Digital Economy Papers 49*. OECD Publishing.

Peru, T. (2022). Digital diplomacy is the next normal. *Diplomatic Courier*. https://www.diplomaticourier.com/posts/digital-diplomacy-is-the-next-normal

Tara, N. (2022). *Africa needs smarter investment in digital infrastructure: Strategies for enticing the private sector. Foresight Africa: Technological innovations: Creating and harnessing tools for improved livelihood*. Foresight Africa.

United Nation Organisation. (2021). *Digital diplomacy in the era of COVID-19*. UN.

Webster, F. (1995). *Theories of the Information Society*. Routledge. doi:10.4324/9780203991367

Chapter 3
COVID-19, Social Media Adoption, and the Future of Digital Diplomacy in Africa

Ranson Sifiso Gwala

https://orcid.org/0000-0002-1545-2259
University of KwaZulu-Natal, South Africa

Pfano Mashau

https://orcid.org/0000-0003-0490-1925
University of KwaZulu-Natal, South Africa

ABSTRACT

The challenges brought by COVID-19 challenged all sectors of society, including public diplomacy. Public diplomacy through the adoption of social media and other digital platforms was transformed into mainstream digital diplomacy. The chapter sought to understand how social media use by diplomats during COVID-19 forced or allowed the introduction of digital diplomacy as a necessary intervention. Using the desktop review of literature, adjoining on systematic literature review was adopted as the research method. The objectives were to investigate how the outbreak of the COVID-19 pandemic affected digital diplomacy in Africa, and what are the implications of the COVID-19 pandemic for the practice of digital diplomacy in Africa. The research identified four themes, these were: the state of African digital diplomacy; social media adoption during COVID-19 in Africa; corona virus and vaccine diplomacy; and anti-corruption public digital diplomacy. The study revealed that African digital diplomacy, however, is closely comparable to its developed countries counterparts.

DOI: 10.4018/978-1-7998-8394-4.ch003

INTRODUCTION

COVID-19 was declared a pandemic by the World Health Organisation, (WHO) in March 2020 after the detection of the spread of COVID-19 cases in many parts of the world. What followed were lockdowns, with countries locking their citizens in their houses and closing all borders of entry (Gwala R.S; Mashau P., 2022). Social media came to the fore with live recordings of small events to the broader audience. Videos calls became another space to connect with family friends and families, and similar platforms for business meetings took off. The relations between nations is natured by advanced diplomacy. Diplomacy infused with latest social media technology brings about digital diplomacy (DD). The study sought to understand the intersection of social media, the impact of COVID-19 and digital diplomacy using technology acceptance model, (TAM). Diplomacy has been disrupted by the digital era. In fact, some scholars have argued that the term "digitalisation of diplomacy" should be used to highlight the implications of ICTs or the use of digital tools for diplomacy, and the practise has been referred to by a variety of names, including e-diplomacy, digi-plomacy, twi-plomacy, networked diplomacy, mobile diplomacy, virtual diplomacy, digital public diplomacy, cyber diplomacy, and digital diplomacy is the most common.

Digital diplomacy scholars contend that when social media is utilized skilfully, there is a propensity for these institutions to reach a wider public, establish new networks, define agendas, and develop relationships with the online public (Willers, 2022; Zaharna, 2022). The worldwide media serve as diplomatic instruments for foreign governments since they disseminate a range of messages and information that could influence public perception of those nations (Saliu, 2020; Sheludiakova et al., 2021). Despite the fact that traditional diplomats and other international stakeholders, such as business executives and representatives of civil society, engage in digital diplomacy in Africa, the practice is still in its infancy compared to other continents. The use of ICTs in foreign policy on the continent is also frequently accidental rather than deliberate and systematic, which means that digital diplomatic practices are not always seen as digital diplomacy in the real sense (Wekesa et al., 2021). As a result, one of the goals of this topic is to give knowledge that will stimulate not only the practical adoption of digital diplomacy but also the creation of digital diplomacy policies and strategies among African foreign policy practitioners.

There is a large research gap on the subject because the study of digital diplomacy is practically non-existent in African colleges (Endong, 2020). Few scholars may be categorized as experts in this topic because there is a dearth of African theory due to a lack of scholarship, African diplomats do not receive professional training in diplomacy and concerns surrounding digital diplomacy in the continent are not being problematized and debated (Arceneaux, 2019). Additionally, in a time of ICT dominance, which marks the Fourth Industrial Revolution, researchers of diplomacy can ill-afford this lack of understanding (4IR) (DANIEL & JAMES, 2020). Africa has historically been on the perimeter of the global knowledge hierarchy, notably when it comes to the regulation of digital technology. The practice of digital diplomacy especially during and after COVID-19 has changed how diplomats feel about digital diplomacy. The training that took place was *adhoc* and inconsistent with a country and between counties within the African continent. The adoption of social media has not become an option, but to what extent it should be used (Gwala, R. S., & Mashau, P., 2023).

Considering the above, research gaps concern:

a)   The need for the unification of traditional diplomacy and extension to digital diplomacy using contemporary theories of technology adoption into a more coherent approach.
b)   The need for African countries to improve digital diplomacy capacity to realise potential. As such, the research problem statement is as follows:

Digital diplomacy development initiatives are built on questionable assumptions and they are on a growth and adoption trajectory (Aggestam et al., 2022; Hedling & Bremberg, 2021)and are not building the necessary skills to improve the digital diplomacy leadership capacity required in African countries to compete and achieve diplomatic outcomes needed by African countries to negotiate with their counterparts anywhere in the world (Huda & Muchatuta, 2022).

THE THEORETICAL FRAMEWORK

The literature review borrows from the Theory of Planned Behaviour, (TPB), Technology Acceptance Model (TAM), and the Unified Theory of Acceptance and Use of Technology (UTAUT) to understand and explain how technology is adopted by organisations and individuals alike. Scanning the emergence of COVID-19 and how it impacted business, governments and diplomats. Further probing how this led to the adoption of technology in the form of social media marketing and other forms of brought by the present era of the fourth industrial revolution (4IR). The adoption of technology has been organically being infused to diplomacy engagements overtime.

The Emergence of COVID-19 In The World

The World Health Organization (WHO) was informed by the China Health Authority on December 31, 2019, the first COVID-19 fatality was reported on January 11, 2020 (Fang et al., 2020). On February 11, 2020, the World Health Organisation (WHO) declared Corona Virus, commonly known as COVID-19, a health emergency. It is a respiratory condition that has an effect on the person's general health (Andrews et al., 2020; Sohrabi et al., 2020). In China, the first COVID-19 case was first noted in December 2019. The new Corona Virus was identified by the WHO as a pandemic disease in March 2020, indicating that it is dispersing quickly over the world's nations. This virus causes fever, cough, sore throat, and breathing difficulties. The COVID-19 pandemic has created risks for economies and business operations, with customers stopping, reducing, or postponing purchases, thereby affecting supply chains and resulting in difficulties in sourcing alternative suppliers (Drydakis, 2022), which affects international trade and international relations between nations. The social distancing was adopted as the first measure to arrest the COVID-19; this was followed by hard lockdowns. International, interprovincial and intercity travel followed suit as measures were strengthened (Cope et al., 2022; Durojaye, 2022).

The Theory of Planned Behaviour (TPB)

The Theory of Planned Behaviour (TPB) was created by (Ajzen, 1985) through expanding the Theory of Reasoned Action (TRA) by adding a new construct called perceived behavioural control (PBC) to account for circumstances in which a person lacks significant control over the targeted behaviour (Tanhan & Young, 2022). A suggestion was made that, in addition to attitudes, use, arbitrary standards, and perceived behavioural controls such the abilities, opportunities, and resources required. The Theory of Planned Behaviour is one of the most significant theories in behavioural science models for forecasting behavioural intentions and actions, and it has been thoroughly validated (Si et al., 2020). The TPB model comes to the conclusion that an individual will be expected to generate the intention to do that action when the opportunity arises if they sense an appropriate level of behavioural control over their behaviour. The three TPB components' direct effects on intention have been studied in several studies. The study's findings supported the importance of the three TPB components and suggested that intention serves as an important stand-in for explaining pro-environmental behaviour (Liu et al., 2021). Theoretical coherence is still lacking, though. Critics have asserted that, under certain circumstances, the attitude had little discernible effect on intention (Abrahamse & Steg, 2011).

Figure 1. Theory of planned behaviour
Source (Ajzen, 1985)

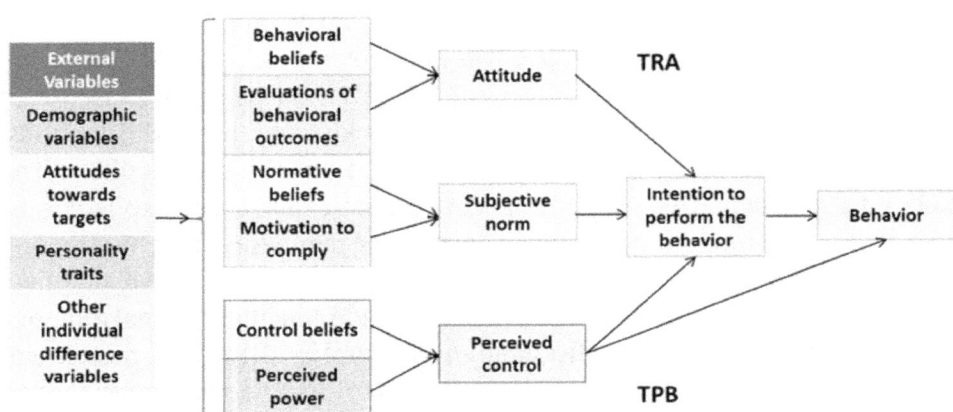

The Technology Acceptance Model, (TAM)

There has been a significant quantity of research on the technology acceptance model (TAM) since it first appeared more than 25 years ago, which clearly supports the model's appeal in the field. The TAM model, which has its origins in psychological theories of reasoned action and planned behaviour, has evolved into an essential tool for understanding the variables that affect whether people will accept or reject new technology. The paper's main objectives are to provide a comprehensive, thoroughly researched resource of past and present references to literature related to TAM as well as to recommend new areas of research for TAM in the future

Figure 2. The Technology Acceptance Model (TAM)
Source: (Davis, 1985)

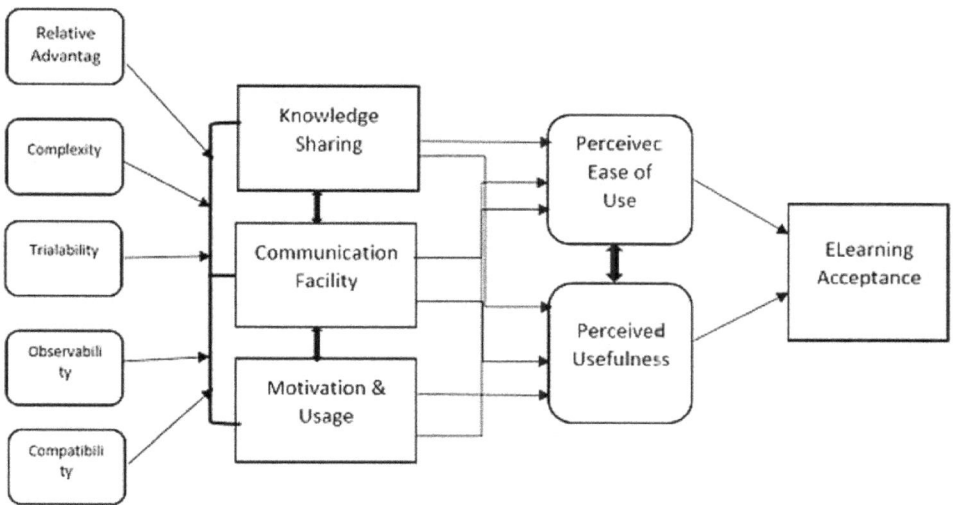

The Technology Acceptance Model (TAM) was suggested by (Davis, 1985). Given the growing popularity of social media, it is crucial to theorise and understand user attitudes and usage patterns of social media websites like Twitter, Facebook, Google+, and LinkedIn in order to develop future understandings and integrate these new technologies. To address such studies on the variables influencing social media usage behaviour, the technology acceptance model (TAM) should be reviewed (Rauniar et al., 2014). Usefulness and ease are the two main criteria involved in user adoption of a technology, according to the technology acceptance model (TAM) (Davis, 1989). TAM has been widely used to assess consumer acceptance of general technology, although it has little explanatory value for purposes relating to particular systems (Jansen-Kosterink et al., 2019). Utilising an example of university students and their perception and adoption of technology, a few terms must be defined. According to Davis, the study's focus on the perceived usefulness (PU) factor is appropriate given that undergraduates think utilising a computer system connects them to a local database or the Internet (Shittu et al., 2011). Instead of using the conventional methods, phones and other electronic gadgets help students with their academic activities. In other words, with technology, students can read and investigate information to complete assignments and research more quickly. Another component of the theory that is pertinent to the investigation is perceived ease-of-use (PEOU) (Bhatiasevi & Naglis,

2016). The simplicity with which students can adopt and employ electronic machine devices will facilitate their usage of digital information resources (DIRs), make learning more convenient for them, and have an impact on their academic activities in daily life.

In other words, using technology automatically makes it easier to find information than using more manual methods. Today instead of taking notes, notes are sent to students before the lecture on their laptops, tablets or even smart phones. This save students time, even those who were absent can access notes wherever they are. Technology adoption amongst the youth is generally higher than in adults (Zhong et al., 2021). Due to its relevance to the issue under inquiry, the researcher chose this idea. In South Africa when cell phones initially arrived in the early turn of the 21st century, it was argued that it was for the rich. But as cell phone became popular and reasonably cheaper, poor households stopped using home desktop telephone which needed a monthly rental whether you had used it or not. As part of the expansion strategies mobile containers were sold to would be small businesses, to spread, familiarise and expand technology adoption to the rural and urban poor alike and they have now become obsolete.

Unified Theory of Acceptance and Use of Technology (UTAUT)

The discussion that follows gives a quick overview of the fundamental elements of the UTAUT model, which are believed to represent the factors that determine IT usage intentions and/or behaviour based on (Venkatesh et al., 2003) research. We start by considering the TRA for the main theoretical foundations of the UTAUT model. The TRA is regarded as one of the most important and fundamental ideas regarding human behaviour (Dwivedi et al., 2019). The TRA's two main constructs are attitudes toward behaviour and arbitrary norms. The TAM was created to forecast IT adoption and use in the workplace, and it has been widely used with a variety of users and technology. The two key constructs in the TAM are perceived usefulness and perceived ease of use. The most thorough upgrade to the Technology Acceptance Model (TAM) is the Theory of Acceptance and Use of Technology (UTAUT) model, which is referred to as a consolidation of past research elements (Qiao et al., 2021). UTAUT provides a more thorough view of technological acceptability than earlier models. The Theory of Planned Behaviour is expanded upon by the UTAUT model. As a reliable method for predicting behavioural intention, UTAUT is described. The UTAUT provides four key direct variables for predicting behavioural intention to use a technology and actual use of the technology and these are:

a) Effort Expectancy,
b) Social Influence,
c) Performance Expectancy, and
d) Facilitating Conditions

and four moderators (age, gender, experience, and voluntariness) (Abbad, 2021; Dwivedi et al., 2020). For example, relative advantage and complexity, which are concepts related to performance expectancy and effort expectancy, respectively, in Diffusion of Innovation theory and similar to perceived usefulness and perceived ease of use, respectively, in TAM (Chen & Aklikokou, 2020). The UTAUT model combines several theoretical concepts and frameworks, such as the Theory of Reasoned Action (TRA) (Ajzen & Fishbein, 1980), Theory of Planned Behaviour (TPB) (Ajzen, 1985), Social Cognitive Theory (SCT) (Bandura, 1986), Technology Acceptance Model (TAM) (Davis, 1989)), Combined-TAM and TPB (C-TAM-TPB) (Taylor & Todd, 1995).

Digital Diplomacy and Social Media Adoption During COVID-19

Diplomacy is a delicate art of maintaining, understanding, and advancing international relations(Fitzpatrick, 2007). International relations purely depend on diplomacy. It is the accepted procedure by which states define their foreign policy goals and coordinate their attempts to influence the judgments and actions of other governments and peoples by communication, negotiations, and other such actions, other than through war and violence. To put it another way, it refers to the long-standing strategies used by nations to protect specific or broader interests, such as the lowering of tensions within or between them. It serves as the primary tool for putting foreign policy's objectives, plans, and overarching strategies into action.

Digital diplomacy is the use of traditional diplomacy with integration with social media (Anton & Lăcătuș, 2022). There is a change in the manner in which things are done as the digital age slowly creeps in (Pipchenko, 2020). Typically, public diplomacy is thought of as a subset of digital diplomacy. It entails the use of digital technology and social media platforms by states to engage in communication with international publics, typically in an inexpensive way, including Twitter, Facebook, and Weibo (Adesina, 2017). The theories of diplomacy identify the following factors of diplomacy which are always in play in diplomatic engagements. These are:

a) Diplomacy as dependent variable
b) Diplomacy as independent variable
c) Bargaining over divergent and incompatible interests
d) Convergent interest

Refering to this next era of diplomacy as hybrid diplomacy, in which actual and virtual engagements will hopefully merge, strengthen, and empower one another (Bjola & Manor, 2022). The COVID-19 outbreak forced diplomacy to shift online for the first time over two years ago. Diplomats were subject to social isolation and quarantine, and multilateral organisations and ministry of foreign affairs (MFAs) were required to lock their doors as of March 2020. In spite of how abrupt the change came about, the procedure seems to have gone quite smoothly in hindsight (Bjola & Manor, 2022). According to the survey by (Chowdhury et al., 2020; Effendi et al., 2020), SMEs impacted by the COVID-19 problem had a high awareness of social media and a strong inclination to embrace it as a channel for marketing their goods and engaging with consumers.

OBJECTIVE AND METHODOLOGY

The following are objectives that will be accomplished by conducting this research

(a) To gain insight into the threats brought by COVID-19 on diplomacy
(b) To examine the digital reforms implemented by diplomats to deal with the COVID-19 situation.
(c) To understand how the outbreak of the COVID-19 pandemic affected digital diplomacy in Africa.
(d) To identify implications of the COVID-19 pandemic for the practice of digital diplomacy in Africa.
(e) To identify the strategic measures for diplomats and diplomacy in the mitigation against the COVID-19 disaster.

Secondary data from a variety of sources (websites, journals, books, and other e-content) are used in this study to accomplish the aforementioned objectives. The study utilized a desktop survey of the literature. This was mainly using Google Scholar, Web of Science, Scopus, etc as the source of peer-reviewed articles databases from the inception of COVID-19 in 2019 to 2022. The survey of literature with identified terms closely linked to the objectives and the topic of the chapter. The survey of literature revealed emergent themes which topics discussed in the next section.

EFFECTS OF THE COVID-19 PANDEMIC ON THE CONDUCT OF PUBLIC DIPLOMACY IN AFRICA

The normal flow of diplomatic business was unexpectedly stopped by the COVID-19. Public diplomacy was not expected to be significantly harmed by COVID-19 despite the slow adoption of digital diplomacy. The COVID-19 examples have shown that digital diplomacy cannot be seen as a replacement for traditional public diplomacy; rather, it must find methods to complement digital diplomacy in order to strengthen diplomatic ties. The emerging themes are:

a)   The state of African digital diplomacy
b)   Social media adoption during COVID-19 in Africa
c)   Corona virus and vaccine diplomacy
d)   Anti-corruption public digital diplomacy

The State of African Digital Diplomacy

Daryl Copeland, an international relations expert, noted that Singapore and Hong Kong had developed internet presences in the late 1990s. Before 2009, American, British, and Swedish diplomats were blogging, while the US State Department had established an office of e-diplomacy and an intranet application called Diplopedia. Other nations with virtual embassies included the Maldives, Sweden, the Philippines, Estonia, Serbia, Colombia, Macedonia, and Albania. Around the same time, the Foreign and Commonwealth Office of the UK established an interactive website connected to YouTube and Flickr. Libya is the number one country with internet penetration at 94,8% as outlined on Table1, with Nigeria positioned fourth at 73% and South Africa positioned at number 13 at 57.5% and lastly at number 58 is Western Sahara at 4.8%.

The average of internet penetration is at 43% and only 21 countries of 58 are above 50% internet penetration. This highlights the greatest challenges that a lot needs to be done going forward.

Table 1. Africa 2022 population and internet users statistics

AFRICA 2022 POPULATION AND INTERNET USERS STATISTICS						
AFRICA TOP 15		**Population** (2022 Est.)	**Internet Users** 31-Dec-00	**Internet Users** 31-Dec-21	**Internet Penetration**	**Internet Growth %** 2000 -21	**Facebook Subscribers** 30-Apr-22
1.	Libya	7,024,811	10	6,658,900	94.8%	58,47%	6,658,900
2.	Kenya	55,752,020	200	46,870,422	85.2%	23,34%	12,445,700
3.	Algeria	45,150,879	50	37,836,425	83.8%	50,76%	26,291,400
4.	Nigeria	211,400,708	200	154,301,195	73.0%	101,48%	31,860,000
5.	Mauritius	1,273,433	87	919	72.2%	956%	919
6.	Seychelles	98,908	6	71,3	72.1%	1,09%	71,3
7.	Morocco	37,344,795	100	25,589,581	68.5%	25,49%	21,730,000
8.	Tunisia	11,935,766	100	8,170,000	68.4%	8,07%	8,170,000
9.	Reunion (FR)	901,686	130	608	67.4%	367%	608
10.	Cabo Verde	565,751	8	352,12	62.3%	4,30%	321
11.	Gabon	2,313,754	15	1,367,641	60.0%	9,02%	872,6
12.	Mali	20,855,735	18,8	12,480,176	59.8%	66,28%	2,033,300
13.	South Africa	60,041,994	2,400,000	34,545,165	57.5%	1,34%	24,600,000
14.	Senegal	17,196,301	40	9,749,527	56.7%	24,27%	3,802,000
15.	Eswatini	1,179,737	10	665,245	56.4%	6,55%	421,5
	TOTAL AFRICA	1,373,486,514	4,514,400	590,296,163	43.0%	12,98%	255,412,900
	Rest of World	6,502,279,070	356,471,092	4,463,594,959	68.6%	88.3%	2,475,026,941
	WORLD TOTAL	7,875,765,584	360,985,492	5,053,891,122	64.2%	100.0%	2,730,439,841

Internet World Stats Source (www.internetworldstats.com.)
Source (www.internetworldstats.com)

This demonstrates that digital diplomacy is not a wholly recent development. When South Africa's Department of International Relations and Cooperation debuted its Internet-streaming broadcaster, Ubuntu Radio, in 2013, it also caught up with similar developments. There is without a doubt enormous potential for the adoption of digital diplomacy in Africa, notwithstanding the infrastructure constraints and other shortcomings the continent faces.

Figure 3 shows that the internet penetration in the world is skewed towards those counties which are well developed. Africa is the least at 43.2% internet penetration, followed by Asia at a difference of over 24% at 67% and Europe and North America are the highest at 89,2% and 93,4% respectively. This is a clear indication of the correlation between continent's economic development and internet penetration. By both nominal GDP and continental GDP per capita, North America leads the world. The nominal per capita GDP of North America is $46,160, which represents 376% of the global average. Oceania ($44,741) comes closely behind North America. Third-ranked Europe ($31,589) and fourth-ranked Asia ($8,034) are separated by a significant amount. South America is in fourth place, Oceania is in third, and Europe is in second. The world's poorest continent is Africa. Africa, South America, and Asia all have lower GDP per capita than the world as a whole. The richest sub-region is Northern America, followed by Australia and New Zealand, Northern Europe, and Western Europe, with four sub-regions having numbers above $50,000. This shows a clear pattern and correlation between internet access and economic growth as per the continental GDP.

The digital divide in digital diplomacy may be closing, according to several studies that show diplomatic bodies in numerous African nations are just as nimble in the digital media realms as their European and American counterparts. Africa with a population of over 1.3 billion people has a penetration rate of just over 43%. This is due to a number of challenges related to access to affordable data, network, broadband and fibre network. Even in places there is access, unreliable electricity due to power outages reduces access and penetration.

Channels of so-called "aid diplomacy" were opened up by the COVID-19 epidemic. A subset of public diplomacy that emphasises the provision of help as one of its main techniques is what is meant by the term "aid diplomacy." People realised that this international problem could not be resolved individually by a single country as a result of the virus's spread at the end of 2019 and in the first half of 2020. South Africa took a lead in ensuring that the developed world donated a substantial number of vaccines to the African continent. This it did along the African Union. The COVID-19 changed how nations conduct public diplomacy. The embassies were closed and trade had to continue. A lot of the work related to travel documentation was conducted online. At least major work was done when the people arrive at the host country. This was unprecedented but it superseded the need for trade and it had to happen. Many nations have continued to learn to improve their digital diplomacy. South Africa is now collaborating with other nations

Figure 3. Internet world penetration rates in Africa (2022)
Source *(www.internetworldstats.com/stats.htm)*

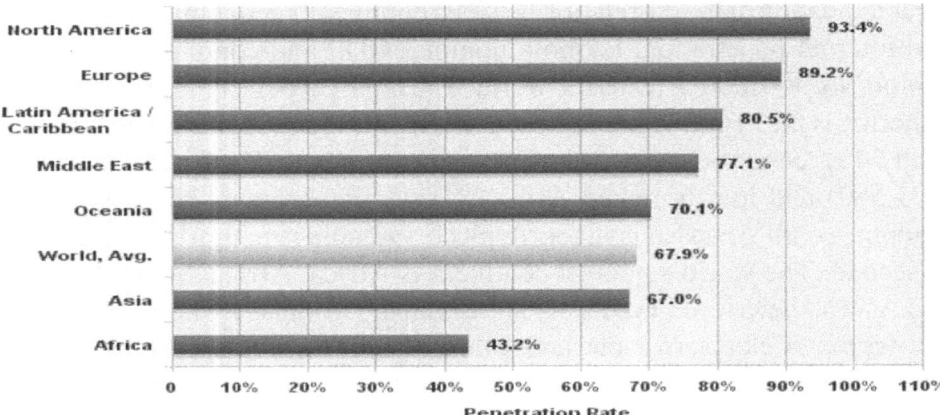

like Kenya who receive their visas at the port of entry as South Africans have been treated for some time. The easing and improvement of ties improves diplomacy and strengthens how diplomacy can advance the work of the diplomatic corps. The transfer of money between countries through platforms like M-PESA (Ndung'u, 2018), that allows for the digital transfer of money between countries has improved people to people relations. Kenya, South Africa, Zimbabwe, Zambia and a number of other countries citizens continue to work in neighbouring countries use M-PESA platform to send money back home, saving traveling money and sustaining their loved one's livelihood. This platform, like any digital platform is susceptible to abuse and crime transfer of cash between countries, but platforms are continuously being improved to counter-cyber-crimes.

Africa has embraced digital diplomacy through these virtual procedures during this chaotic period. For instance, despite the pandemic, African governments, the African Union (AU), and non-governmental groups have organised a number of online conferences on peace and security that have gathered thousands of African stakeholders (Bradshaw et al., 2021). A three-week long online seminar called *"Silencing the Guns"* was successfully held by the AU in May 2020. Both in-person and online attendees participated, adding to the discussion and forming new relationships. Since the start of the pandemic, African leaders have held online meetings with stakeholders, this was adopted by a number of African presidents and diplomatic corps (Turianskyi & Wekesa, 2021).

Along with lowering the cost of bringing together important parties, these online talks also made it possible for more people to participate at once, which sped up decision-making (Banerjee, 2019). The Central African Republic, South Sudan, Somalia, Mali, and the Darfur region of Sudan have all seen an improvement in the peace and security as a result of cooperation between the UN Security Council (SC) and the AU Peace and Security Council (Martinsson & Thillberg, 2020). It is essential that this partnership be formalised and that the issue of how to pay for AU peacekeeping missions be addressed and resolved. Currently in its second year as an elected non-permanent member of the UNSC is South Africa. Throughout

South Africa's time in office, they have fought for inclusive discourse and the peaceful resolution of conflicts in order to advance global peace and security (Khalid et al., 2021). South Africa public diplomacy has continued to be peace and loyalty in the world and being non-partisan. The South African constitution has become its public diplomacy as it promotes fairness and equality amongst its citizens, although South Africa is the most unequal country in the world (Adetiba, 2021; Ballard et al., 2005).

Social Media Diplomacy Adoption During COVID-19 in Africa

Public diplomacy (PD) refers to the methods used by different nations to build channels of contact with their target audiences (Thiel, 2022). In order to control the foreign policy environment, these state communication initiatives target audiences from other nations and cultures. The emergence of social media upended this hierarchy of communication since it permitted more direct, honest, and fair communication between diplomatic corps and nations (McGregor, 2020). According to Holmes, (Bjola & Holmes, 2015), digital diplomacy is the use of digital information and communication technologies, including the Internet, to further diplomatic goals (Adesina, 2017).The technological capability of Digital Diplomacy (DD) allows for the facilitation of the development of relationships on social media (e.g., following on Twitter, likes on Facebook). Second, DD largely relies on user-generated content, including reviews, comments, and videos (Nantongo, 2019). During COVID-19 lockdowns, communication between nations was mostly conducted via television and digital channels. One of the tasks carried out by the government was the collection of citizens from all over the world (Oksana Andriivna et al., 2021). While other nations were quick to locate and retrieve their vital personnel from their countries of origin in order to ensure that they were prepared to battle COVID-19, this was a clear example of diplomacy in action (Medinilla et al., 2020). Social media was mostly used to identify these citizens because global travel had already been restricted. Granting African presidents used Facebook at a differing rate, posts were mainly focused on diplomatic ties. Using contingency analysis, the majority of diplomatic relations posts occurred in April, the same as posts on information and orientation (Adikpo, 2022).

By the end of 2021, Statistics Internet World Stats implies that despite holding the continent's highest population and internet users, Facebook users in Nigeria ranked less than Egypt, which ranks third in population and internet penetration by population respectively. Ethiopia whose population ranks second maintains the fourth position in internet penetration, internet, and Facebook users respectively. South Africa and DR Congo whose population is fourth and fifth among the countries selected ranked third and last for internet and Facebook users respectively. Given the recent theorizing of social media impact during the coronavirus pandemic particularly strong use patterns, purposes and impacts are expected in the use of Facebook by the African leaders. Today, from Kenya and South Africa through Mali and Namibia to Rwanda, digital diplomacy has become a received idea in the African continent. Twiplomacy notes for instance that the leaders of Eritrea, Mauritania and Swaziland are not on Facebook. The same survey revealed that only few African leaders (notably Paul Kagame of Rwanda, Nana Akufo-Ado of Ghana and Uhuru Kenyatta of Kenya are either very active on social media and/or have millions of followers on such networking sites as Facebook, Twitter and Instagram.

The continent was responsible for 53% of social media restrictions in 2021, with targeted apps including WhatsApp, Facebook Messenger, Facebook, Twitter and Instagram. The road to use social media by diplomats as well as Presidents in Africa will take some time. In Nigeria, Ghana and in Zimbabwe there are times when the internet is blocked in order to limit the spread of information especially during elections and elections campaigns (Chibuwe, 2020), this was either as a deterrent to fake media or a weapon against the opposition (Conroy-Krutz, 2020). President – Mokgweetsi Masisi, President of Botswana has a very active page which communicates both diplomatic, social and economic issues. President PatriceTalon of Benin also hosts an active twitter account and very active in utilising the medium for communication. A number of African countries have greatly improved in having a footprint on social media, like Egypt Spokesperson, posts all updates on twitter. The most notable African countries which are lagging behind are very few, and those countries tend to see social media as a hindrance in their governance issues or their portrayal to the outside world.

Coronavirus and Vaccine Diplomacy

As China, India, Russia, and the United States compete to project influence through their home-grown vaccines and the signing of vaccine purchase agreements with countries who had less access to vaccines, COVID-19 vaccines had emerged as a new public diplomacy tool. According to (Kickbusch & Liu, 2022; Shakeel et al., 2019), vaccine diplomacy is a field of global health diplomacy that encourages the use and administration of vaccines in order to advance both common foreign policy goals and bigger global health goals (Kickbusch & Liu, 2022). China began a public diplomacy campaign in April 2020 to position itself as a pioneer in global health by sending masks, medical teams, and test kits abroad following the successful domestic COVID-19 mitigation (Lee, 2021; Lee & Kim, 2021). The United States and Russia collaborated to combat polio during the Cold War. One of the first nations to use medical diplomacy as a tool for foreign policy was Cuba (Bhattacharya et al., 2021). Africa remained as the pariah of the world which only received vaccine donations, even when Africa wanted to buy, it was on the basis of vaccine excess from the vaccine producing countries and their governments (Rogerson & Rogerson, 2020). The spread of COVID-19 in Africa surprised many, as it did not kill many people given the limited resources that Africa could use for COVID-19 mitigation. South Africa established a Pfizer and Aspen vaccine manufacturing in Port Elizabeth, in the Eastern Cape Province, and in Cape Town, Western Cape respectively, but all ingredients were imported from the United States of America (Lamptey et al., 2022). The infancy of the African drug manufacturing industry showed the over-reliance on multinational companies which are well funded in their countries of origin to continue to act in advancement of vaccine diplomacy, now and in the future. Africa remains an incapable consumer whose industries are at infancy or are yet to be developed. Africa is yet to develop any substantial influence on vaccines let alone vaccine diplomacy (Kararach, 2022; Shah, 2021).

Anti-Corruption Public Digital Diplomacy

Numerous academics have declared the connection between e-governance and the fight against corruption. Information and communication technology is regarded as a crucial tool that can be used to combat the despicable corruption trend (Subroto et al., 2021). Information and Communications Technology (ICT) can reduce corruption by encouraging good governance and keeping an eye on both the actions of the government and the governed (Adam & Fazekas, 2021). Among other measures used by the government in many developing countries around the world, the use of electronic measures in daily governmental activity has a significant impact on the battle against corruption (Kelly et al., 2022). Different regimes have made a number of attempts to stop corruption. Two distinct themes in African corruption emerge: first, the participation of high-ranking government officials in grand corruption; and second, the illegal and covert movement of looted properties and monies outside of the continent. As a result, when corruption is committed, African States experience a significant drain on their resources (Tucker, 2022). Foreign nations risk becoming safe havens for stolen African assets, which causes Africa's growth to stall, be paralysed, and be hijacked by the offenders (Schmidt, 2018). Compounded by a long list of issues like weak institutions, weak laws, a culture of impunity, a lack of the rule of law, a growing wealth gap, the pains of underdevelopment, understaffed and underfunded anti-corruption authorities, and the simple lack of a strong political will, exacerbate this problem (Fomunung, 2018). The South African government has adopted the National Anti-Corruption Strategy 2020-2030 as well as the Local government Anti-corruption strategy (Maluleke et al., 2022). Kenya has also adopted similar strategies known as National Ethics and Anti-Corruption Policy – (EACC). Nigeria and Zambia have also followed suit following the African Union (AU) Summit on combating corruption (Kelly, 2020).

IMPLICATIONS FOR THE FUTURE OF PUBLIC DIPLOMACY IN AFRICA

Social media sites like Twitter, Facebook, WhatsApp, and Instagram have shown to be effective tools for influencing the public, particularly in terms of strengthening a nation's reputation, among many other purposes. For instance, a large number of African leaders, MFAs, and other relevant organisations own social media accounts, particularly Twitter and Facebook. Notably, President Muhammadu Buhari of Nigeria has more than 5 million followers on Facebook, Instagram, and Twitter combined, making him the most followed African leader. They have made serious comments on their social media which have given serious indications on how they think and to what direction they are likely to lean.

The South African President, Mr Cyril Ramaphosa and the former Finance Minister, Mr. Trevor Manuel have cited people who have written to them on social media giving them advice or seeking help. This shows that public diplomacy is gradually learning and including digital diplomacy. The relations between nations is closely being shown on social media. People are likely to make decisions based on what they see on social media (Boulianne, 2020). South Africa, Nigeria, and other nations have shared xenophobic videos against African refugees (Gwala R.S. & Mashau P, 2022), this went viral on social media sending wrong public diplomacy between African nations (Okeke, 2022). Social media including fake social media continue to influence how people think, it is therefore critical that countries communicate and monitor fake social media in order to counteract fake news that is counter-productive in the eyes of individuals who take time to understand the clear and correct position of their neighbouring nations.

The internet adoption is relatively slow in Africa at lower than 50%. The data continues to amongst the most expensive in the world. The issues of broadband and spectrum continue to influence the slow rollout pace of access to many remote countries (Forge & Vu, 2020). Yet, Africa has not lagged behind in digital diplomacy adoption. The nations continue to understand that digital diplomacy is critical in promoting their countries to their neighbouring countries and the world at large. Africa largely has vast amounts of land, which facilitates and promotes African beauty and international tourism. Showcasing African amazing rivers, mountains, waterfalls, fauna and flora represents the extent to which digital diplomacy can play a role in packaging Africa as a tourist friendly continent (Huber, 2020; Mushawemhuka, 2021). The perception of Africa can be gradually changed by digital diplomacy.

The well planned digital diplomacy has a greater potential of teaching the western world that Africa is now just about the wild life, it's about its rich culture, its traditions and can teach the world alternative ways of life rooted in ubuntu culture and leadership (Gray et al., 2022; Xiang & Leung, 2022).

CONCLUSION

The use of technology in society spiked during COVID-19, with many of the users showing clearly their lack of understanding of the use of these platforms (Gwala, R. S.; Mashau, P., 2022). Although WhatsApp, Twitter, Facebook, Instagram, and many other popular social media platforms have been about for some time, but they have been gaining popularity amongst the youth and they were not regarded as mainstream means of marketing of communicating formal business messages. The growth of popularity of these platforms and the growing number of people following superstars and influencers has changed the landscape how business looks at social media. The adoption of social media by diplomats gave rise to digital diplomacy. The study will seek to understand how social media adoption has been shaping and developing what is now termed digital diplomacy. This shall relate to how the emergence of COVID-19 necessitated and fastened the technology infusion and adoption into mainstream diplomacy. Society in general has not understood what public diplomacy is. Public diplomacy has also been affected by the COVID-19 pandemic. During COVID-19 diplomacy also underwent changes and confirmed that digital diplomacy could no longer be delayed. Digital diplomacy became the norm in communicating and rescuing African citizens who were all over the world who needed to get back home. The unity of Africa was also emphasised when the continent pulled together in the fight against COVID-19. The injustice of poverty in Africa was also laid bare, but African continent pulled together to access aid vaccines in the face of vaccine diplomacy and vaccine dominancy by the rich nations. Whilst some rich nations advanced their vaccine diplomacy, Africa struggled to keep their nations' economies running. This meant that leading economies in Africa along with the African Union had the hardest task to use public and digital diplomacy to advance the needs of Africa to access vaccines, medicines and food. The challenges of Africa continue to be uneven development, poverty, unemployment and wars in very few countries. The energy just transition is also another dilemma facing Africa. Whilst Africa continues to be underdeveloped with vast deposits of coal, it is now forced to adopt policies that it cannot afford. The developed world builds their economies using coal and fossil fuels, polluting the world, and now Africa must be the first to pay the price without reaping the rewards of development and growth. Public diplomacy should take into account public justice into the world development and fight against public inequity.

REFERENCES

Abbad, M. M. (2021). Using the UTAUT model to understand students' usage of e-learning systems in developing countries. *Education and Information Technologies*, 26(6), 7205–7224. doi:10.100710639-021-10573-5 PMID:34025204

Abrahamse, W., & Steg, L. (2011). Factors related to household energy use and intention to reduce it: The role of psychological and socio-demographic variables. *Human Ecology Review*, 30–40. https://www.jstor.org/stable/24707684

Adam, I., & Fazekas, M. (2021). Are emerging technologies helping win the fight against corruption? A review of the state of evidence. *Information Economics and Policy*, 57, 100950. doi:10.1016/j.infoecopol.2021.100950

Adesina, O. S. (2017). Foreign policy in an era of digital diplomacy. *Cogent Social Sciences*, 3(1), 1297175. doi:10.1080/23311886.2017.1297175

Adetiba, T. C. (2021). Public Diplomacy and South Africa's Response to Xenophobia. *African Renaissance*, 18(3), 59.

Adikpo, J. A. (2022). Adoption of Social Media during COVID-19 Pandemic by African Presidents: A Cross-Sectional Study of Selected Facebook Accounts. *The Journal of Communication Inquiry*, 0(0), 01968599221144314.

Aggestam, K., Rosamond, A. B., & Hedling, E. (2022). Feminist digital diplomacy and foreign policy change in Sweden. *Place Branding and Public Diplomacy*, 18(4), 314–324. doi:10.105741254-021-00225-3

Ajzen, I. (1985). From intentions to actions: A theory of planned behavior. In *Action control* (pp. 11–39). Springer., doi:10.1007/978-3-642-69746-3_2

Ajzen, I., & Fishbein, M. (1980). Understanding Attitudes and Predicting Social Behavior. Englewood Cliffs, NJ: Prentice- Hall.

Andrews, M., Areekal, B., Rajesh, K., Krishnan, J., Suryakala, R., Krishnan, B., Muraly, C., & Santhosh, P. (2020). First confirmed case of COVID-19 infection in India: A case report. *The Indian Journal of Medical Research, 151*(5), 490. doi:10.4103/ijmr.IJMR_2131_20 PMID:32611918

Anton, A., & Lăcătuş, M. (2022). Digital Diplomacy: The Case of the Embassy of Sweden in Bucharest. In *Diplomacy, Organisations and Citizens* (pp. 199-218). Springer.

Arceneaux, P. C. (2019). *Information Intervention and the Need for a Social Cybersecurity Perspective: The Power Struggle between Digital Diplomacy and Computational Propaganda*. University of Florida.

Ballard, R., Habib, A., Valodia, I., & Zuern, E. (2005). Globalization, marginalization and contemporary social movements in South Africa. *African Affairs, 104*(417), 615–634. doi:10.1093/afraf/adi069

Bandura, A. (1986). *Social foundations of thought and action*.

Banerjee, A. (2019). Blockchain with IOT: Applications and use cases for a new paradigm of supply chain driving efficiency and cost. In *Advances in computers* (*Vol. 115*, pp. 259-292). Elsevier.

Bhatiasevi, V., & Naglis, M. (2016). Investigating the structural relationship for the determinants of cloud computing adoption in education. *Education and Information Technologies, 21*(5), 1197–1223. doi:10.100710639-015-9376-6

Bhattacharya, S., Saleem, S. M., Shikha, D., Gokdemir, O., & Mehta, K. (2021). Role of vaccine science diplomacy in low-middle-income countries for eradicating the vaccine-preventable diseases: Targeting the "LAST MILE". *Journal of Family Medicine and Primary Care, 10*(8), 2739. doi:10.4103/jfmpc.jfmpc_2253_20 PMID:34660398

Bjola, C., & Holmes, M. (2015). *Digital Diplomacy*. Taylor & Francis. doi:10.4324/9781315730844

Bjola, C., & Manor, I. (2022). The rise of hybrid diplomacy: From digital adaptation to digital adoption. *International Affairs, 98*(2), 471–491. doi:10.1093/ia/iiac005

Boulianne, S. (2020). Twenty years of digital media effects on civic and political participation. *Communication Research, 47*(7), 947–966. doi:10.1177/0093650218808186

Bradshaw, S., Campbell-Smith, U., Henle, A., Perini, A., Shalev, S., Bailey, H., & Howard, P. N. (2021). *Country case studies industrialized disinformation: 2020 global inventory of organized social media manipulation.* Oxford Internet Institute.

Chen, L., & Aklikokou, A. K. (2020). Determinants of E-government adoption: Testing the mediating effects of perceived usefulness and perceived ease of use. *International Journal of Public Administration 43*(10), 850–865. doi: 10.1080/01900692.2019.1660989

Chibuwe, A. (2020). Social media and elections in Zimbabwe: Twitter war between Pro-ZANU-PF and Pro-MDC-A Netizens. *Communicatio: South African Journal of Communication Theory and Research, 46*(4), 7–30. doi: 10.1080/02500167.2020.1723663

Chowdhury, M., Sarkar, A., Paul, S. K., & Moktadir, M. (2020). A case study on strategies to deal with the impacts of COVID-19 pandemic in the food and beverage industry. *Operations Management Research*, 1-13.

Conroy-Krutz, J. (2020). The squeeze on African media freedom. *Journal of Democracy, 31*(2), 96–109. doi:10.1353/jod.2020.0024

Cope, K., Somin, I., & Stremitzer, A. (2022). Vaccine Passports as a Constitutional Right. *Ariz. St. LJ, 54*, 25. https://www.research-collection. ethz.ch/bitstream/handle/20.500.11850/504018/CLE_WP_2021_10. pdf?sequence=1

Daniel. E. E., & James, H. M. (2020). Narratives and Industrial Policy. The Oxford Handbook of Industrial Policy, 284.

Davis, F. D. (1985). *A technology acceptance model for empirically testing new end-user information systems: Theory and results.* Massachusetts Institute of Technology. https://www.researchgate.net/profile/ Sonam-Mathur-3/publication/301824711_Demographic_Influences_on_ Technology_Adoption_BehaviorA_Study_of_E-Banking_Services_in_India/ links/5aec0c02458515f59981f28c/Demographic-Influences-on-Technology-Adoption-BehaviorA-Study-of-E-Banking-Services-in-India.pdf

Davis, F. D. (1989). Perceived Usefulness, Perceived Ease of Use, and User Acceptance of Information Technology. *Management Information Systems Quarterly, 13*(3), 319–340. doi:10.2307/249008

Drydakis, N. (2022). Artificial Intelligence and reduced SMEs' business risks. A dynamic capabilities analysis during the COVID-19 pandemic. *Information Systems Frontiers*, 1–25. PMID:35261558

Durojaye, E. (2022). Between a rock and a hard place:(un) balancing the public health interventions and human rights protection in the COVID 19 era in South Africa. *International Journal of Human Rights*, 26(2), 332–347. doi:10.1080/13642987.2021.1926238

Dwivedi, Y. K., Rana, N. P., Jeyaraj, A., Clement, M., & Williams, M. D. (2019). Re-examining the unified theory of acceptance and use of technology (UTAUT): Towards a revised theoretical model. *Information Systems Frontiers*, 21(3), 719–734. doi:10.100710796-017-9774-y

Dwivedi, Y. K., Rana, N. P., Tamilmani, K., & Raman, R. (2020). A meta-analysis based modified unified theory of acceptance and use of technology (meta-UTAUT): A review of emerging literature. *Current Opinion in Psychology*, 36, 13–18. doi:10.1016/j.copsyc.2020.03.008 PMID:32339928

Effendi, M. I., Sugandini, D., & Istanto, Y. (2020). Social media adoption in SMEs impacted by COVID-19: The TOE model. *The Journal of Asian Finance, Economics and Business, 7*(11), 915-925.

Endong, F. P. C. (2020). Digitization of African Public Diplomacy: Issues, Challenges and Opportunities. *Infonomics Society, 1*, 9.

Fang, H., Wang, L., & Yang, Y. (2020). Human mobility restrictions and the spread of the novel coronavirus (2019-nCoV) in China. *Journal of Public Economics, 191*, 104272. doi:10.1016/j.jpubeco.2020.104272 PMID:33518827

Fitzpatrick, K. (2007). Advancing the new public diplomacy: A public relations perspective. *The Hague Journal of Diplomacy*, 2(3), 187–211. doi:10.1163/187119007X240497

Fomunung, J. S. (2018). *Africa's Path to Economic Development: A Guide for Policy Makers and Scholars*. Spears Media Press.

Forge, S., & Vu, K. (2020). Forming a 5G strategy for developing countries: A note for policy makers. *Telecommunications Policy*, 44(7), 101975. doi:10.1016/j.telpol.2020.101975

Gray, D. L., Ali, J. N., McElveen, T. L., & Sealy, M. (2022). The Cultural Significance of "We-Ness": Motivationally Influential Practices Rooted in a Scholarly Agenda on Black Education. *Educational Psychology Review*, *34*(4), 1–29. doi:10.100710648-022-09708-y

Gwala, R. S., Mashau, P. (2022). Covid-19 and the future of migration and mobility in Africa: A systematic literature review, Journal of Nation-building & Policy Studies (JoNPS), Special Issue, 223-255. DOI: https://doi.org/doi:10.31920/2516-3132/2022/s1s1a12

Gwala, R. S., & Mashau, P. (2023). COVID-19 and SME Adoption of Social Media in Developing Economies in Africa. In S. Qalati, D. Ostic, & R. Bansal (Eds.), *Strengthening SME Performance Through Social Media Adoption and Usage* (pp. 133–152). IGI Global., doi:10.4018/978-1-6684-5770-2.ch008

Gwala, R. S., & Mashau, P. (2022). Corporate governance and its impact on organisational performance in the Fourth Industrial Revolution: A systematic literature review. *Corporate Governance and Organizational Behaviour Review*, *6*(1), 98–114. doi:10.22495/cgobrv6i1p7

Hedling, E., & Bremberg, N. (2021). Practice approaches to the digital transformations of diplomacy: Toward a new research agenda. *International Studies Review*, *23*(4), 1595–1618. doi:10.1093/isr/viab027

Huber, M. (2020). *Developing Heritage–Developing Countries: Ethiopian Nation-Building and the Origins of UNESCO World Heritage, 1960–1980* (Vol. 1). Walter de Gruyter GmbH & Co KG. doi:10.1515/9783110681017

Huda, M. I. M., & Muchatuta, E. T. (2022). African diplomacy issues and challenges. *Journal of Positive School Psychology*, 2137–2146-2137–2146.

Jansen-Kosterink, S., Dekker-van Weering, M., & van Velsen, L. (2019). Patient acceptance of a telemedicine service for rehabilitation care: A focus group study. *International Journal of Medical Informatics*, *125*, 22–29. doi:10.1016/j.ijmedinf.2019.01.011 PMID:30914177

Kararach, G. A. (2022). *Disruptions and rhetoric in African development policy*. Routledge. doi:10.4324/9781003153467

Kelly, R. M. (2020). *Policies and strategies in Kenya's response to the war on terror: a critical evaluation*. Strathmore University.

Kelly, S. J., Derrington, S., & Star, S. (2022). Governance challenges in esports: A best practice framework for addressing integrity and wellbeing issues. *International Journal of Sport Policy and Politics*, *14*(1), 151–168. doi:10.1080/19406940.2021.1976812

Khalid, N., Izzi, V., Bishop, V., MacNeil, C., Altiok, A., Onyango, W., Atuhaire, G., Klugman, J., Moore, M., & Budur, D. (2021). *Securitizing Youth: Young People's Roles in the Global Peace and Security Agenda*. Rutgers University Press.

Kickbusch, I., & Liu, A. (2022). Global health diplomacy—Reconstructing power and governance. *Lancet*, *399*(10341), 2156–2166. doi:10.1016/S0140-6736(22)00583-9 PMID:35594877

Lamptey, E., Senkyire, E. K., Benita, D. A., & Boakye, E. O. (2022). COVID-19 vaccines development in Africa: A review of current situation and existing challenges of vaccine production. *Clinical and Experimental Vaccine Research*, *11*(1), 82. doi:10.7774/cevr.2022.11.1.82 PMID:35223668

Lee, S. T. (2021). Vaccine diplomacy: Nation branding and China's COVID-19 soft power play. *Place Branding and Public Diplomacy*, 1–15.

Lee, S. T., & Kim, H. S. (2021). Nation branding in the COVID-19 era: South Korea's pandemic public diplomacy. *Place Branding and Public Diplomacy*, *17*(4), 382–396. doi:10.105741254-020-00189-w

Liu, Y., Shi, H., Li, Y., & Amin, A. (2021). Factors influencing Chinese residents' post-pandemic outbound travel intentions: an extended theory of planned behavior model based on the perception of COVID-19. *Tourism Review*.

Maluleke, A., Edoun, E. I., & Pooe, S. (2022). Education as an analysis of Poverty Status of Households in Limpopo, South Africa. [IJEB]. *International Journal of Economic Behavior*, *12*(1), 83–100.

Martinsson, P., & Thillberg, E. (2020). *AU-led Peace Operations: The Case of the AMISOM KDF's Local Peacebuilding Engagement in Southern Somalia*. Jubbaland Region.

McGregor, S. C. (2020). "Taking the temperature of the room" how political campaigns use social media to understand and represent public opinion. *Public Opinion Quarterly*, *84*(S1), 236–256. doi:10.1093/poq/nfaa012

Medinilla, A., Byiers, B., & Apiko, P. (2020). African regional responses to COVID-19. *ECDPM, DP, 2*, 272.

Mushawemhuka, W. J. (2021). *A comprehensive assessment of climate change threats and adaptation of nature based tourism in Zimbabwe*. University of Johannesburg.

Nantongo, S. K. (2019). *A Comparative Analysis of Digital Diplomacy by the Obama Administration to the Trump Administration and Its Influence On Effective US Foreign Policy*. United States International University-Africa.

Ndung'u, N. (2018). The M-Pesa technological revolution for financial services in Kenya: A platform for financial inclusion. In *Handbook of blockchain, digital finance, and inclusion* (Vol. 1, pp. 37–56). Elsevier. doi:10.1016/B978-0-12-810441-5.00003-8

Okeke, K. N. (2022). The Psychology and the Psychosocial Impacts of Xenophobia and Nativism. *Interrogating Xenophobia and Nativism in Twenty-First-Century Africa*, 37.

Oksana Andriivna, B., Olena Vasylivna, K., Lopushanskyy, V., Valeriia Mykhaylivna, S., & Yukhymets, S. (2021). Psychological difficulties during the covid lockdown: Video in blended digital teaching language, literature, and culture. *Arab World English Journal (AWEJ) Special Issue on Covid, 19*.

Pipchenko, N. (2020). Digital diplomacy: How international actors transform their foreign policy activity. *Ukraine Analytica*, 02(20), 19–25.

Qiao, P., Zhu, X., Guo, Y., Sun, Y., & Qin, C. (2021). The development and adoption of online learning in pre-and post-COVID-19: Combination of technological system evolution theory and unified theory of acceptance and use of technology. *Journal of Risk and Financial Management*, *14*(4), 162. doi:10.3390/jrfm14040162

Rauniar, R., Rawski, G., Yang, J., & Johnson, B. (2014). Technology acceptance model (TAM) and social media usage: An empirical study on Facebook. *Journal of Enterprise Information Management*, *27*(1), 6–30. doi:10.1108/JEIM-04-2012-0011

Rogerson, C. M., & Rogerson, J. M. (2020). COVID-19 tourism impacts in South Africa: Government and industry responses. *Geo Journal of Tourism and Geosites*, *31*(3), 1083–1091. doi:10.30892/gtg.31321-544

Saliu, H. (2020). The Evolution of the Concept of Public Diplomacy from the Perspective of Communication Stakeholders. *Medijska istraživanja: znanstveno-stručni časopis za novinarstvo i medije, 26*(1), 69-86.

Schmidt, E. (2018). *Foreign intervention in Africa after the cold war: Sovereignty, responsibility, and the war on terror.* Ohio University Press.

Shah, R. J. (2021). The COVID Charter: A New Development Model for a World in Crisis. *Foreign Affairs, 100,* 179.

Shakeel, S. I., Brown, M., Sethi, S., & Mackey, T. K. (2019). Achieving the end game: Employing "vaccine diplomacy" to eradicate polio in Pakistan. *BMC Public Health, 19*(1), 1–8. doi:10.118612889-019-6393-1 PMID:30654797

Sheludiakova, N., Mamurov, B., Maksymova, I., Slyusarenko, K., & Yegorova, I. (2021). Communicating the Foreign Policy Strategy: on Instruments and Means of Ministry of Foreign Affairs of Ukraine. *SHS Web of Conferences.* SHS. 10.1051hsconf/202110002005

Shittu, A. T., Basha, K. M., AbdulRahman, N. S. N., & Ahmad, T. B. T. (2011). Investigating students' attitude and intention to use social software in higher institution of learning in Malaysia. *Multicultural Education & Technology Journal.* doi:10.1051/shsconf/202110002005

Si, H., Shi, J., Tang, D., Wu, G., & Lan, J. (2020). Understanding intention and behavior toward sustainable usage of bike sharing by extending the theory of planned behavior. *Resources, Conservation and Recycling, 152,* 104513. doi:10.1016/j.resconrec.2019.104513

Sohrabi, C., Alsafi, Z., O'neill, N., Khan, M., Kerwan, A., Al-Jabir, A., Iosifidis, C., & Agha, R. (2020). World Health Organization declares global emergency: A review of the 2019 novel coronavirus (COVID-19). *International Journal of Surgery, 76,* 71–76. doi:10.1016/j.ijsu.2020.02.034 PMID:32112977

Subroto, G., Bari, A., & Pakendek, A. (2021). Indonesian White-Collar Crime; The Social Psychology Approach And The Need Of Conspiracy Theory To Deal With It. *British Journal of Criminology, Law & Justice, 1*(2), 122–141.

Tanhan, A., & Young, J. S. (2022). Muslims and mental health services: A concept map and a theoretical framework. *Journal of Religion and Health, 61*(1), 23–63. doi:10.100710943-021-01324-4 PMID:34241742

Taylor, S., & Todd, P. A. (1995). Understanding information technology usage: A test of competing models. *Information Systems Research, 6*(2), 144–176. doi:10.1287/isre.6.2.144

Thiel, M. (2022). EU public diplomacy in the United States: Socio-political challenges & EU delegation agency. *Journal of Contemporary European Studies*, 1–13.

Tucker, O. M. (2022). *The Flow of Illicit Funds: A Case Study Approach to Anti–Money Laundering Compliance*. Georgetown University Press. doi:10.2307/j.ctv2m2fv8m

Turianskyi, Y., & Wekesa, B. (2021). African digital diplomacy: Emergence, evolution, and the future. *South African Journal of International Affairs, 28*(3), 341–359. doi:10.1080/10220461.2021.1954546

Venkatesh, V., Morris, M. G., Davis, G. B., & Davis, F. D. (2003). User acceptance of information technology: Toward a unified view. *Management Information Systems Quarterly, 27*(3), 425–478. doi:10.2307/30036540

Wekesa, B., Turianskyi, Y., & Ayodele, O. (2021). *Introduction to the special issue: Digital diplomacy in Africa* (Vol. 28). Taylor & Francis.

Willers, J. O. (2022). Seeding the cloud: Consultancy services in the nascent field of cyber capacity building. *Public Administration, 100*(3), 538–553. doi:10.1111/padm.12773

Xiang, Y., & Leung, C. T.-L. (2022). The utilisation of Ubuntu across cultures: A case study of a rural development programme in China. *International Social Work*, 00208728221129364. doi:10.1177/00208728221129364

Zaharna, R. (2022). *Boundary Spanners of Humanity: Three Logics of Communications and Public Diplomacy for Global Collaboration*. Oxford University Press. doi:10.1093/oso/9780190930271.001.0001

Zhong, Y., Oh, S., & Moon, H. C. (2021). Service transformation under industry 4.0: Investigating acceptance of facial recognition payment through an extended technology acceptance model. *Technology in Society, 64*, 101515. doi:10.1016/j.techsoc.2020.101515

Chapter 4
Digitalization of Diplomacy:
Defense Diplomacy in Turkey

Ekrem Yaşar Akçay
Hakkari University, Turkey

Murat Mutlu
Hakkari University, Turkey

ABSTRACT

This chapter will discuss the changes and developments in defense diplomacy in Türkiye. Türkiye, which has taken very important steps in digitalization with the COVID-19 process, has also taken important steps in defense and digitalization of defense. The study consists of two chapters. In the first chapter, the concept of defense diplomacy will be discussed and the transformation of defense diplomacy in the historical process will be examined. In the second part, defense diplomacy practices in Türkiye and the digital transformation of defense diplomacy will be examined with examples. A descriptive method will be used in the study.

DOI: 10.4018/978-1-7998-8394-4.ch004

INTRODUCTION

Diplomacy, defined as the peaceful conduct and maintenance of relations between states by officials appointed by states, is one of the important tools that states put into practice to realize their security and foreign policy objectives and strategies (Abdurahmanlı, 2021, p. 581). Diplomacy, which is a concept as deep-rooted and old as the history of humanity, has been a method applied by states in the international system throughout history. Initially, diplomacy was used by states as a temporary method such as making agreements and announcing peace conditions, but as time passed and the international conjuncture changed, diplomacy also changed. In particular, states without military power have used diplomacy to survive. Diplomatic moves made temporarily have become the constant presence of representatives in the states concerned. States have opened representative offices in the relevant countries and appointed officials who will stay here permanently. This has also changed the scope of diplomacy. Diplomatic negotiations to make agreements and give information included gathering information and conducting intelligence activities (Cooper, Heine & Thakur, 2013, p. 3).

As time passed and the international conjuncture changed, diplomacy, which began to be applied in various forms such as coercive diplomacy, permanent diplomacy, summit diplomacy, and shuttle diplomacy, began to become more important than military power in realizing the foreign policies of states (Nicolson, 1941, p. 27). In this sense, the hard power used to define military power has been replaced by soft power with the increase in the effect of diplomacy. This situation has also led to a change in the scope of power. Over time, not only military but also economic and technological power, which is one of the elements of soft power, has increased its influence. In particular, with the increase in the influence of public diplomacy, which is used to define the method used by the government of one country to influence and direct the citizens of another country in line with their national interests and ideologies, a bond that cannot be broken between soft power and diplomacy has been formed (Cull, 2009, p. 17).

Considering the developments in technology in recent years, almost everything from education to health, from foreign policy to economy has started to become digital. Almost all applications and all activities have started to be done easily and conveniently in a very short time using digital platforms. Diplomacy has also had its share of this situation and diplomacy has taken on a new form (Cull, 2008, p. 33). Digital diplomacy, also defined as the implementation of public diplomacy on digital platforms, has gained an important place in the foreign policy practices of states (Adesine & Summer, 2017, p. 4). Because of the restrictions and quarantine practices implemented with the COVID-19 pandemic, governments have carried out their activities online. State leaders attended the meetings online and tried to influence societies by explaining their policies through social media.

The digitalization of diplomacy has also been reflected in the policies implemented by states in many areas such as defense and security. For example, defense diplomacy, which is examined as a branch of both soft power and public diplomacy, which is a new concept theoretically but has a long history in practice, has also come under the influence of digitalization. This study discusses the historical development of defense diplomacy, the transformation it has experienced in this process, and how much it has been affected by the digitalization process. As a case study, defense diplomacy practices in Türkiye were examined. The study consists of two chapters. In the first chapter, the definition, characteristics, and historical development of defense diplomacy are described. In the second chapter, the implementation and development process of defense diplomacy in Türkiye is explained. A descriptive method was used in the study. Because of the lack of sufficient studies on defense diplomacy, it is hoped that your work will make an important contribution both in terms of theory and case studies.

DEFENSE DIPLOMACY AND DIGITALIZATION OF DEFENSE DIPLOMACY

Defense diplomacy is a concept that refers to the achievement of foreign policy aims through the peaceful use of the military and security instruments of the state. Defense diplomacy is used as a term that encompasses everything from officer exchanges, training missions, arms transfers, and strategic assistance to joint military exercises designed to improve interoperability between national units (Leman & Jardine, 2021, p. 2). Defense diplomacy is a method that has existed in various forms for hundreds of years. Because defense diplomacy is very important for states The idea that many of the international problems can best be addressed through interaction between military personnel and contact between the armed forces, rather than negotiations between diplomats, political leaders, or others, is of interest.

Although the concept of defense diplomacy was used by Western countries for the first time, the conduct and implementation of defense diplomacy was in no way limited to Western countries. Defense diplomacy has been used by actors in the international system. The concepts of defense diplomacy and military diplomacy, which are used in the same sense, are different from each other. Military diplomacy is recognized as a subset of defense diplomacy (Singh, 2021, p. 109). However, it is also necessary to distinguish defense diplomacy from "strategic" or "security" diplomacy. Strategic or security diplomacy encompasses all diplomatic efforts designed to support or protect a country's strategic or security interests and can therefore be undertaken in any way. "Strategic" or "security" diplomacy is about questions about goals, not means. Defense diplomacy, on the other hand, is more about the means used than the ends pursued. While most defense diplomacy is for strategic or security purposes, what makes it defense diplomacy is conducted through military channels (White, 2014, p. 10).

Defense diplomacy, based on the non-violent use of military forces and means, is the combined use of diplomatic and military means. Defense diplomacy is the conduct of negotiations and other relations between nations, armies of nations, and citizens of nations by military diplomats. Defense diplomacy, which also means that military personnel carry out and maintain diplomatic activities to prevent international conflicts and resolve disputes between states, has a mission that also serves the democratic development of the armed forces (Cottey & Forster, 2004, p. 6). Defense diplomacy is a concept based on cooperation between military forces in times of peace, aiming to ensure, protect and maintain peace and describes mutual interaction (Plessis, 2008, p. 92).

Defense Diplomacy is an integral part of foreign policy and state security, helps to increase military cooperation between countries, and strengthens relations between states. In the pre-Cold War period, defense diplomacy was known as military cooperation with other states to counter an enemy, while in the post-Cold War period, cooperating with potential enemies, serving with them, and supporting good governance and human rights were defined as ensuring democracy. In other words, the objectives of the armed forces within a state such as defense, deterrence, and intervention have undergone changes in the post-cold war period and the scope of the duties of the armed forces has expanded (Midhio & Priyono, 2019, p. 64).

Defense diplomacy is a concurrent of soft power. The use of soft power instead of hard power in the international system has made defense diplomacy an important strategic tool in realizing the foreign policies of states. The exchange of officers between states, exercises, assistance to the armies of states in peacetime, training, and consultancy activities have increased the impact of soft power. For this reason, defense diplomacy is considered part of soft power. For example, Joseph Nye Soft Power, who introduced the concept of *soft power in his work Soft Power: The Means to Success in World Politics*, divided power into three categories as military power, economic power, and soft power and included defense diplomacy under the title of soft power (Nye, 2005, p. 18- 21). Similarly, Gregory Winger, in his work *The Velvet Gauntlet: A Theory of Defense Diplomacy*, divided power into three as hard power, economic power, and soft power, and examined defense diplomacy under soft power (Winger, 2014, p. 4). A state can apply its soft power in two different ways: direct and indirect. The indirect use of soft power is ensured through public diplomacy. For example, states carry out educational activities and programs to influence the citizens of other states and to get their support. In this way, they indirectly use soft power. The direct use of soft power is in the form of conducting international conferences, conducting intergovernmental visit programs, conducting interstate military exercises, establishing direct contact with governments, and conducting military training between the relevant states, and in general, the direct use of soft power is realized through defense diplomacy (Balcı, 2018, p. (49).

Although defense diplomacy is considered a new concept, it has a long history. Since the 17th century, states have sent their military personnel to other countries to gather information, conduct observations, and communicate with allies. The military personnel involved have used defense diplomacy in carrying out their duties. For example, Cardinal Richelieu was one of the first to use and implement the concept of defense diplomacy. Richelieu, who was an important politician as well as a holy joe, sent the officers in France to Sweden and the Netherlands to observe, follow and gather information on military developments during the Thirty Years' War between 1618 and 1648 (Nathan, 1993, p. 637). Later changes in the international system led to the development of both diplomacy and defense diplomacy. From the 18th century onwards, defensive attachés were assigned to embassies, and many countries began to assign these attachés to establish colonial empires from the beginning of the 19th century (Grina, 2017, p. 155). At the same time, during this period, states used defense diplomacy to intervene in and prevent crises, and to observe and collect information about other states (Lamsal, 2022, p. 83). In the 20th century, in 1927, the Imperial Defense College was established in England, and the officers trained there were sent to train the members of the British Commonwealth. With the establishment and operation of the college, a common defense doctrine was developed in the United Kingdom. (Defense Academy of the United Kingdom, 2009).

Immediately after the end of the Cold War, the Eastern European countries worried the Western countries. Western governments feared that the large Soviet-style armies of the former Warsaw Pact countries would be major obstacles to the delicate transition to democracy, and that unreformed armies would derail the transition process. Because of this fear, Western governments have adopted a strategy that uses their military power to help restructure the armed forces of the former Warsaw Pact countries. Through nonviolent activities such as officer exchange and training programs, Western governments have mobilized their defense institutions to reform Eastern Europe's militaries (Mearsheimer, 1990, p. 8). For example, in the early 1990s, Germany provided defense assistance to Russia, which was in East Germany and wanted to withdraw. Until 1994, the German Ministry of Defense defined "military-political" cooperation with the former Warsaw Pact countries as one of the main missions of its armed forces. But in 1994, it became clear that traditional security perspectives were expanding and that the role of the military went beyond deterrence and defense, and the German Defense White Paper was published, incorporating a series of measures into the mission of the armed forces that would include the resolution of conflicts without gaining a military dimension (Resdal, 1994). In 1994, NATO launched the Partnership for the Peace process to build democracy, ensure political stability, cooperate, conduct training, conduct exercises, and modernize their armies in newly independent states after the dissolution of the USSR (NATO, 2020).

In addition, the United States has established a form of defense diplomacy. The American form of defense diplomacy was part of a broader concept of defense engagement that encompassed two groups cf military activities. The first group was Foreign Military Interaction, which included military aid, military training, joint planning, exercises, and operations, and the second group was defense diplomacy. In the American context, this was an unofficial term used to describe the military and defense support of foreign policy in peacetime. The difference between the two groups was that defense diplomacy activities often took place in a policy vacuum, were undeveloped, unfunded, and were not based on legislation, unlike Foreign Military Interaction activities (Department of the Army, 2001, p. 8).

As can be seen, although some countries have engaged in a series of military cooperation and aid activities since the 1990s, they have not used the concept of direct defense diplomacy. The term defense diplomacy was first used by Defense Secretary George Robertson in the UK Strategic Defense Assessment White Paper (Grattan, 2011, p. 101). This document which was published in 1998, was the first comprehensive review of defense policy pursued by the

British government since 1993 and took into account developments in the post-Cold War world, such as the sprouting of international peace-support operations. The Strategic Defense Assessment White Paper represented a reassessment of Britain's security interests and defense needs and set out the roles, missions, and capabilities of the UK armed forces to "meet these new realities" (House of Commons, 1998). Unlike previous policies, the defense diplomacy mission defines a new role for defense in promoting understanding and trust between all European powers and assisting the development of modern democratic armed forces, particularly in central and Eastern Europe. To help explain the concept, the UK Department of Defense has published a special supplementary document on Defense Diplomacy as part of the Strategic Defense Assessment White Paper. In the document, the Ministry of Defense acknowledged that although the idea of defense diplomacy was not new, the duties and powers of military forces were extended to a wider area to help prevent the escalation or emergence of conflicts (House of Commons, 1998).

When all these evaluations are considered, it is seen that until recently, defense diplomacy was examined from a partial and narrow point of view, but it is understood that more holistic and comprehensive examinations have been carried out recently. As changes took place in many areas of the international conjuncture, including social, political, and technological, states began to make reforms to keep up with these changes, and these reforms were reflected in almost every field. This is also evident in defense diplomacy. While hard power elements such as military cooperation and bilateral defense agreements were used in defense diplomacy in the pre-Cold War period, the scope of defense diplomacy changed and expanded when new elements threatening the security of states emerged in the post-Cold War period. To ensure cooperation between senior military and civilian officials, to conclude defense cooperation agreements, to train foreign military and civilian personnel, to exchange military personnel, to provide military assistance, to visit military elements, to conduct exercises, to appoint defense attachés to foreign countries, to ensure peace-building in foreign countries, to ensure the establishment and development of democracy in foreign countries, and to ensure and maintain political stability in foreign countries and to maintain defense It is counted among the duties of diplomacy (Winger, 2014, p. 3). Moreover, in the new era, defense diplomacy has often been associated with conflict prevention and security sector reform. For example, the Commander of the Air Force of the Republic of Singapore, Major General Ng Chee Khern, said that the state seeks to develop mutually beneficial relations with friendly countries and armed forces to contribute to a stable international and regional environment through defense diplomacy (Khern, 2009, p. 3).

Defense diplomacy, by contrast, exists with a conceptual ambiguity that prevents the further study of the issue. Because defense diplomacy was first used to cover a range of pre-existing activities, it was never developed as a conceptually distinct idea. Therefore, defense diplomacy has become an expression that has no fixed meaning just like the concept of terrorism, and has become only a tiny part of conceptual coherence, and without conceptual boundaries, it has become impossible to say what constitutes an act of defense diplomacy. This situation has caused defense diplomacy to lose its meaning. Therefore, Because of the lack of a universally accepted definition of defense diplomacy, states have adapted the content of the concept to the wants and needs of their security and foreign policies. After all, over time, defense diplomacy has been one of the guiding mechanisms used to help the West confront the current global security environment (Koerner, 2006, p. 3). It has become an increasingly important part of states' government strategies. For example, in the United Kingdom, defense diplomacy has become one of the eight "defense missions" of the military (Muniruzzaman, 2020, p. 66).

In addition, the extent to which defense diplomacy can succeed requires more comprehensive negotiations. But defense diplomacy has not been able to prevent conflicts between states. Ultimately, it is not the lack of cooperation between the two militaries that leads to conflict, but rather the national objectives, threat perceptions, and often irreconcilable differences in ideology that fuel disagreements and wars. For example, pre-war military cooperation between the Soviet Union and Nazi Germany in the 1930s, could not prevent a bloody war between the two states, which lasted nearly four years, including the training of officers. Therefore, states should set their expectations regarding defense diplomacy from a realistic perspective (Gambhir, 2021).

With the developments in technology, digitalization has shown its effect in almost every field of society in social life. This situation has started to allow the digitalization of defense diplomacy. Although digitalization, which has started to show its effect, especially with the COVID-19 pandemic, shows itself in diplomacy in general, it is still quite new in the field of defense diplomacy. On the other hand, it is possible to carry the ideas about digitalization in the field of defense and defense diplomacy back to the 1980s. For example, Colonel Jack Thorpe, who served in the US Air Force, came up with an idea in the 1980s that envisaged simulator networks for combat planning and exercises, and that over time the conditions of war would be inextricably more similar to the real ones (Thorpe, 2010). With this idea, the U.S. Defense Advanced Research Projects Agency (DARPA) established SIMNET, and in 1987 it started to use it for training purposes by fielding it. In the following periods, synthetic training environments were established and it was aimed to conduct live, virtual training. In this way, it is aimed to make military training portable and available wherever it is needed (Miller, 2015). Today, this process is intended to be supported by the military metaverse. Although the COVID-19 pandemic process has adversely affected the whole world, it has left a positive mark in terms of experiencing these developments.

The digitalization process in defense diplomacy is also realized through computer games. Thanks to advanced gaming technologies, real life is recreated in a virtual environment. For example, it is possible to see this situation in games such as Company of Heroes and Counter-Strike. Such games have been an important strategy for the interaction of the British army with societies. For example, in Army Esports, real soldiers can be asked questions as well as people are communicated with. In this way, it facilitates communication and interaction between units (STM Thinktech, 2022).

Since the beginning of the COVID-19 pandemic, digitalization has shown its effect in almost every field in society like a spider's web, and has caused several new developments. For example, the US has established a virtual battlefield with the Synthetic Training Environment that can simulate cities in North Korea, making it easy and costly for the US to conduct military training and exercises in this environment. Especially with the COVID-19 pandemic, military exercises have started to be held online. In April 2022, the United States and South Korea conducted a nine-day digital military exercise. This exercise was conducted through computer simulation (Sputnik, 2022). On February 25, 2022, the NATO Extraordinary Online Summit was held. Many countries participated online in the meeting on Russia's aggression against Ukraine (NATO, 2022).

DEFENSE DIPLOMACY IN TÜRKİYE

Defense Diplomacy Practices in Türkiye

Through joint military training, exercises, meetings, sports, exchange programs, defense attachés, peacekeeping missions, grants within the scope of the defense industry, technology transfer, and transfer projects, Türkiye is trying to use defense diplomacy first in its region and then in different continents. Türkiye's military structure is conducive to this situation. According to a study conducted in England, Türkiye has the second most active army abroad after the US. Türkiye currently has a military presence in 12 countries. It also has military bases in some of these countries. The largest military base abroad is located in Somalia, a strategically important point. Located in the capital Mogadishu, this base provides training support to the Somali army. Apart from the United Nations (UN) and NATO missions, Türkiye has deployed troops in Qatar, Somalia, Albania, Libya, and Azerbaijan under bilateral agreements in Iraq and Syria on its initiatives. Türkiye is currently contributing to the peace-support operation in the territorial waters and offshores of Bosnia and Herzegovina, Kosovo, Afghanistan, Lebanon, Iraq, and Somalia (Erenel, 2020).

As of today, Türkiye has a total of 253 missions, including 144 Embassies, 13 Permanent Representations, 94 Consulates General, 1 Consular Agency, and 1 Trade Office (Türkiye Cumhuriyeti Dışişleri Bakanlığı, 2022) Most of the countries where embassies are located have a Defense Attaché (Military Attaché) through the Turkish General Staff. Defense attachés are generally responsible for all bilateral military and defense relations. Today, there are Military Attachés in 83 countries (Milli Savunma Bakanlığı, 2022)

International relations have an important contribution to the sales and marketing of defense products. Defense diplomacy remains important in relations between states and this sector is not adversely affected by any crisis. Because when crises arise, everyone tries to increase security. Therefore, there is a growing trade in terms of all crisis regions of the world and it is seen that crises are great in this geography (Erdoğan, 2022).

Türkiye's vision of becoming one of the world's largest arms exporters is becoming a reality as the country's air system technologies continue to gain international attention. For example, Türkiye exports Baykar's Bayraktar TB2 tactical UAVs and related weapons to the Middle East, Central Asia, and North Africa on a large scale (Aviation Week Network, 2021). A defense and aerospace sector that has increased the indigenousness rate in the defense industry above 80% and reduced its foreign dependency to a minimum level, has been able to bring the share of defense and aerospace products to developed countries in its exports, has been able to produce and export products with high technological added value and thus can contribute to additional employment to its economy, clustered and synergized within itself, has made Türkiye a global power that can produce more independent policies. brings.

Geopolitically, Türkiye has shown how drones can be a strong foreign policy presence. This technology has helped the country push its geopolitical rivals and conduct military operations that could once be very costly and dangerous. As European countries struggle to forge partnerships in the Middle East and North Africa, and the US returns to Asia, Türkiye has shown how investment in drone capabilities can prevent its partners from seeking new allies (European Council on Foreign Relations, 2021).

Turkish troops are currently present in at least nine countries, from Iraq to Somalia to Northern Cyprus, and operate a large base in Qatar. Türkiye also controls parts of the territory in northern Syria following a series of cross-border operations. In the field of diplomats, the axis of the military and the intelligence community coincides with the expansion of Türkiye's armed forces, which are already NATO's second-largest armed forces. The government has doubled military spending over the past decade and plans to produce all arms hits by 2023 (Yackley, 2020).

The unique defense industry products and the increasing level of deterrence of the Turkish Armed Forces (TAF) have contributed greatly to the success of the actions of Turkish foreign policy. To maintain this position, Türkiye needs to develop its arms industry and increase its share in the global arms trade. The increased level of deterrence of the TAF has contributed greatly to the success of the actions of Turkish foreign policy. To maintain this position, Türkiye needs to develop its arms industry and increase its share in the global arms trade. The TAF's increasing level of deterrence of the TAF has contributed greatly to the success of the actions of Turkish foreign policy. To maintain this position, Türkiye needs to develop its arms industry and increase its share in the global arms trade (Metin, 2022).

As a NATO member, Türkiye has an interest in maintaining its power and dominance in the Middle East North Africa region (Middle East North Africa (MENA). Türkiye is considered a country that has calculated power at the global level. In the last decade of political influence and economic potential, Türkiye has been trying to catch up with international relations that have been delayed for decades. Türkiye's role in the face of the Middle East crisis is getting stronger. Türkiye's strong commitment to supporting international security has been successful in its cooperation with Indonesia, meaning it has added opportunities to market developers in Europe and the Middle East Region. For this reason, diplomacy studies are seen as inevitable in terms of defense, thus proving that the development of the military industry is a game changer (Balcı, 2018, p. 55).

Türkiye uses its arms sales not only to make money but also to build alliances and establish a patronage network, especially on the African continent. In this sense, Türkiye cooperates with certain regions of the world.

For example, the TAF contributes to peace-support operations in Bosnia and Herzegovina, Kosovo, Afghanistan, Lebanon, and Somalia. In particular, Türkiye carried out its activities within the scope of the first post-Cold War defense diplomacy in Bosnia and Herzegovina. The training given to the army of Bosnia and Herzegovina is aimed at increasing the capabilities of the personnel in various fields such as communication and interoperability, especially technology, rather than military training. In this respect, the TAF personnel become a promotional tool reflecting the level of development of the Turkish army and Türkiye and gain respect (Balcı, 2018, p. 53).

Kosovo is another country where Türkiye uses its soft power effectively within the scope of defense diplomacy activities. TAF personnel also served as part of the Kosovo "Turkish Delegation Presidency" in the city of Prizren, where the Muslim population is densely populated. In Kosovo, the people generally see Türkiye as a brotherly country. One point identified by personnel serving in Kosovo shows the contribution of the role played by the TAF in the 1999 NATO air campaign, which was the use of military force, to Türkiye's image. This is because Turkish soldiers played the most active role in the Albanians' 1999 NATO intervention, and planes of other countries did not deliberately hit Serbian targets, but Turkish jets destroyed Serbian targets with great success and saved themselves (Doğan, 2014, p. 79).

Türkiye's third peacekeeping mission, which continues with certain intensities until the Taliban comes back to power, is taking place in Afghanistan. In 2001, the International Relief Force (ISAF) was established by the resolution of the UN Security Council and took office in 2002. Türkiye, ISAF in Afghanistan; Working with the civilian population on important issues such as education and health services, improving urban water, and ensuring and supporting the establishment of administrative administration, infrastructure, and agricultural problems, the TAF elements carry out various tasks ranging from training and advising Afghan security forces in Afghanistan to ensuring security within the country or the security of aid sent to the country. But these duties, Because of Türkiye's national restrictions, are outside the scope of responsibility and do not include mine clearance, counter-terrorism, and anti-drug activities (Doğan, 2014, p. 80).

Türkiye, which has an intensive defense diplomacy relationship with Iraq, is the country with the most military bases in Iraq It carries out operations against the PKK terrorist organization in northern Iraq. Thanks to bilateral agreements signed with the Kurdistan Regional Government of Iraq, Türkiye maintains dozens of military bases in and around Bashiqa by training Peshmerga and Sunni Arab fighters. The central Iraqi government opposes Türkiye's military operations on Iraqi soil (Aksoy, 2021).

Digitalization Process and Applications of Defense Diplomacy

Technical development called "Web 2.0"; allows internet users to take part in social networks and to conduct communication experiences through these network interactions regardless of the source or destination. The foreign policy objectives and priorities determined by the political subject are increasingly supported in the context of public diplomacy, which is constructed through strategic communication efforts coordinated by the relevant institutions of the state. Social networks are the most important communication channels where this support is embodied. Digital communication platforms, which offer highly appropriate opportunities in terms of open dialogue and mutual interaction, are changing the armies of the world, which are known as extremely rigid institutions in terms of their extreme sensitivity to information security and their protectionist approaches (Akaydın Aydın and Sadakaoğlu, 2020, p. 225).

This orientation of the new public diplomacy is also manifested in practices and similar to propaganda, one-way information flow is not adopted. Developments in new communication technologies offer opportunities where people can come together and connect as if they were face-to-face. These opportunities offer public diplomacy the opportunity to talk and discuss with much smaller budgets instead of cultural activities that require huge financial resources (Yağmurlu, 2019, p. 1289).

Türkiye's defense diplomacy practices have also entered the digital platform, especially with the COVID-19 pandemic. For example, the TAF Partnership for Peace Training Center Command carries out courses, seminars, and Mobile Training Team (MTT) activities within the scope of its main education and training activities. In 2021, a total of 1,046 personnel, 624 of whom were Turkish and 422 foreigners, from 38 different countries participated in 16 courses (8 courses open to the participation of foreigners within the scope of Covid-19 measures were carried out by distance education method.) (Milli Savunma Bakanlığı, 2021, p. 61).

In 2020, aid was provided to Somalia, Montenegro, Rwanda, Kyrgyzstan, and Kosovo. NATO Mission in Iraq (NMI), Operation Kosovo (KFOR), Resolute Support Mission in Afghanistan (until the KDM-Taliban came to power), EUFOR ALTHEA Operation in Bosnia and Herzegovina, Operation Naval Guard in the Mediterranean continued to be followed, coordinated, and directed national contributions. NATO Military Committee Chiefs of Staff Meetings were held on January 15-16, 2020 in Belgium via remote video call on May 14, 2020, and the NATO Military Committee Chiefs of Staff Conference was held on September 18, 2020, via remote video call. The Chiefs of Staff of NATO member countries participated in the meetings and the Conference (Milli Savunma Bakanlığı, 2021, p. 58).

In May-December 2020, to add value to the institution during the pandemic process, the consultancy activity on "Simplification of the Legislation with the Business Continuity Maturity Scale Against the COVID-19 Outbreak and the Use of Technology in this Context" was carried out remotely in the Ministry units (Milli Savunma Bakanlığı, 2021, p. 67).

On August 10-12, 2022, NATO Allied Command Transformation organized an Educational Technology Conference online and physically. The event, to which a limited number of participants were invited according to their competence, was held with more than 100 senior managers from 10 companies. [REMOVED HYPERLINK FIELD] BİTES, which has strong infrastructure, products, and projects in the field of the metaverse, made a presentation on "Taking advantage of developing metaverse technologies to transform NATO's education and training" (TRT Haber, 2022).

With the presentation made by BİTES Senior Manager Uğur Coşkun and Business Development Director İhsan Yusuf Akbuağa, the first presentation of the company's military metaverse technologies infrastructure "MiliVerse" was made at NATO. BİTES has made significant progress on the way to the metaverse with the SahaExpo 2020 Virtual Fair and the SahaExpo 2021 Hybrid Fair, which were held as a first in Türkiye and the defense industry. While SahaExpo 2020 was an application consisting of static and visuals, SahaExpo 2021 was implemented as an application consisting of dynamic and 3D models. Miliverse, which was developed with the virtual and hybrid exhibition system XperExpo infrastructure of BİTES and introduced in NATO, can be configured in line with the needs of users from beginning to end and offers an innovative solution for three-dimensional virtual organizations and corporate requirements. The MiliVerse application has integrated a customizable avatar system so that users can better express themselves in virtual environments. With this system, users can design their digital twins in the virtual environment as they wish. Users can access and interact with each other simultaneously from the current browsers on their computers, mobile phones, tablets, and Oculus Quest VR glasses. While one user enters the application through VR glasses, another user can share the experience and receive training in the same environment by joining the system from his mobile device (TRT Haber, 2022).

The app welcomes users with the main hall and four different virtual environments that can be accessed from this hall. These four virtual environments consist of two different maintenance hangars, BİTES Showroom, and a conference room. For the technology demonstration of MiliVerse, the Atak helicopter, one of the products that also achieved export success in the Turkish defense industry, was selected. The user who wants to access the Atak produced by TAI with the maintenance-repair modeling developed can log in with VR glasses and perform the determined dismantling-tool procedures of his helicopter with directions. Users who log in from other devices can watch the dismantling-tooling procedures live. In the other virtual experience area, the trainer aircraft KT-1T used by the TAF is located in the maintenance hangar. A user who logs in with VR glasses can interact with the aircraft through directions, and users who log in from other devices can follow this experience live. Another scene of the application is the company area of BİTES. Users who visit the company area have the opportunity to examine the company's products in 3D, and access detailed information from documents, images and videos. Users can interact with the products, interact with the product responsible and get detailed information about the product. In the conference room, which has a three-dimensional futuristic design, users have the chance to sit where they want, share in cinema vision and watch the conference live, similar to the real-world experience. From this point of view, it is possible to say that defense diplomacy in Türkiye has started to manifest itself in the digital field (TRT Haber, 2022).

CONCLUSION

Diplomacy, known as the peaceful pursuit of states' foreign policies, has an ancient history and has been an important tool used by states in the international system in their foreign policies. With the changes in the international conjuncture over time, diplomacy has changed both in form and content, and new diplomacy types and methods have been used. For example, today, issues such as tourism, health, and education have become significant tools used by states to realize their foreign policy strategies. States are trying to play a more active role in the international system by using these elements for their foreign policy interests. Especially as globalization and developments in technology make the international system digital and complex day by day, states have begun to use new diplomacy methods and digitalize these models. One of these states was Türkiye. Türkiye has started to use diplomacy as a crucial tool in almost everything from education to health, from defense to trade. Defense diplomacy has also been an important foreign policy tool for Türkiye.

Defense diplomacy, which has a long history, is a new concept used by the states in the international system. Defense diplomacy is a method often used in relations between states throughout history. Although hard power comes to mind primarily when it comes to defense diplomacy, defense diplomacy has been one of the important elements of soft power. Defense diplomacy, which has started to be used by many states in the world, has enabled both the development and diversification of diplomacy. Defense diplomacy, led by states such as the US and the United Kingdom, is also used intensively by Türkiye.

Türkiye's military cooperation with many countries, training, and exercises contributed to the development of defense diplomacy. On the other hand, thanks to the recent developments in technology and the military field in Türkiye, its effectiveness in the field of defense diplomacy has increased. The export of military weapons and equipment produced in Türkiye to other states has also positively affected Türkiye's development in defense diplomacy.

With the recent developments in Türkiye, the Turkish defense industry has turned from being a need-oriented industry to an area where Türkiye will gain commercial, industrial, scientific and technological gains in the long run. Thanks to the national technology move it initiated, Türkiye abandoned its position as a sought-after customer of the world's leading arms producer countries; instead, it has evolved into a rising star of the global defense market. With these developments, Türkiye Baykar Technology has developed the third generation of Armed/Unarmed UAV platforms thanks to the UAVs concept developed by ROKETSAN and TAI. The main reason for the developments in the defense industry in Türkiye has been realized with the opportunities and capabilities of self-defense. Thanks to this move by Türkiye, the Turkish defense industry does not only meet the needs of the Turkish Armed Forces; it designed military and civilian vehicles and equipment to meet the demands of the armies of other countries. The Turkish defense industry has turned from being a need-oriented industry into an area where Türkiye will gain commercial, industrial, scientific and technological gains in the long run. The progress Türkiye has made in the defense industry has not only been a game changer on the battlefield but has also provided a significant opportunity for Turkish diplomacy.

The COVID-19 pandemic, which emerged in China in the last period of 2019, has affected the whole world and many business lines, especially daily work, from education to health, have started to become more digital. Diplomacy has also had its share in this situation. Because face-to-face meetings canceled because of the pandemic, relations between states have been tried to be realized on digital platforms and using social media, and both other states and other societies have been affected by digital platforms. However, these developments in diplomacy are still in the beginning stages of defense diplomacy. Efforts are being made to ensure that defense diplomacy takes place on digital platforms. However, it is not yet in the process of fully implementing. The same is true in Türkiye. Türkiye is also among the states that use defense diplomacy intensively. Türkiye's training, exercises, collaborations, and imports and exports of military equipment to other countries are shown as examples of this situation. With the COVID-19 pandemic, Türkiye, as every country does, carries out its activities related to defense diplomacy on digital platforms. With the miliverse it has developed and introduced in the recent period, it has taken another crucial step in the name of the digitalization of defense diplomacy. Senior officials in Turkey take crucial steps in defense diplomacy by making several meetings, both on the phone and on online platforms. But this situation is still in the preparation stage. In this sense, although defense diplomacy is a part of diplomacy, it has not developed as much as diplomacy. Although significant developments and successful moves have been made in digital diplomacy in Turkey, no significant steps have been taken in defense diplomacy yet. However, it is thought that Turkey will be more successful thanks to the significant steps it will take in the future. Since digitalization is a must in almost every field in today's world, Turkey's digitalization in defense diplomacy will make a crucial contribution to Turkey. In the past, power for states was determined by military power, economic power, population power, the state's army, and weapon capacity, but today power has become a concept related to technology. Both the scope and nature of power have changed. Therefore, the more Turkey attaches importance to digital elements, especially in defense diplomacy, the more it will develop. Because the capacity of digitization for states has become dependent on the capacity of power. However, thanks to the developing technology, it seems that it will not take long for defense diplomacy to develop and digitalize.

ACKNOWLEDGMENT

This research received no specific grant from any funding agency in the public, commercial, or not-for-profit sectors.

REFERENCES

Abdurahmanlı, E. (2021). Definition of diplomacy and types of diplomacy used between states. *Anatolian Academy Social Sciences Journal*, *3*(3), 580–603.

Adesina, O., & Summers, J. (2017). Foreign policy in an era of digital diplomacy. *Cogent Social Sciences*, *3*(1), 1–14. dci:10.1080/23311886.20 17.1297175

Akaydın Aydın, A. ve Sadakaoğlu, M. C. (2020). İdeoloji ve instagram: Türk silahlı kuvvetlerinin yönettiği ınstagram hesabının kamu diplomasisi açısından incelenmesi. *The Turkish Online Journal of Design. Art and Communication*, *10*(3), 221–231.

Aksoy, H. (2021). *Excursus: Turkey's military engagement abroad.* Retrieved August 20, 2021, from https://www.cats-network.eu/topics/visualizing-turkeys-foreign-policy-activism/excursus-turkeys-military-engagement-abroad#c4312

Balcı, A. (2018). Savunma Diplomasisi Kavramı, Özellikleri ve Uygulamaları. *Ufuk Üniversitesi Sosyal Bilimler Enstitüsü Dergisi*, *7*(14), 45–58.

Cooper, A. F., Heine, J., & Thakur, R. (2013). Introduction: The challenges of 21st-century diplomacy. In A. F. Cooper, J. Heine, & R. Thakur (Eds.), *The Oxford Handbook of Modern Diplomacy* (pp. 2–27). Oxford University Press. doi:10.1093/oxfordhb/9780199588862.001.0001

Cottey, A., & Forster, A. (2004). *Reshaping defense diplomacy: New roles for military cooperation and assistance.* Oxford University Press.

Cull, N. J. (2008). Public diplomacy: Taxonomies and histories. *The Annals of the American Academy of Political and Social Science*, *616*(1), 31–54. doi:10.1177/0002716207311952

Cull, N. J. (2009). *Public diplomacy: Lessons from the past.* Figueora Press.

Defense Academy of the United Kingdom. (2009). *A History Of RCDS*. Retrieved July 27, 2022, from https://webarchive.nationalarchives.gov.uk/ukgwa/20091211061225/http://www.da.mod.uk/colleges/rcds/About_Us/A%20History%20of%20RCDS

Department of the Army. (2001). *US Army Field Manual 3-0, Operations*. United States Army.

Doğan, H. (2014). Türk silahlı kuvvetlerinin kamu diplomasisi faaliyetleri. *Güvenlik Bilimleri Dergisi*, *3*(2), 67–90. doi:10.28956/gbd.283039

Erdoğan, A. (2022). *Savunma diplomasisi ve milli güç*. Retrieved March 2 2022, from https://thinktech.stm.com.tr/tr/turk-savunma-sanayiinin-adaptasyon-ve-donusumunde-kuresel-oyuncularla-rekabet

Erenel, F. (2020). *Türkiye'nin savunma diplomasisi ve yurtdışında asker bulundurma*. Retrieved December 8, 2020, from https://www.gazetebirlik.com/yazarlar/turkiyenin-savunma-diplomasisi-ve-yurt-disinda-askeri-varlik-bulundurma/

European Council on Foreign Relations. (2021). *Turkey's drone diplomacy: Lessons for Europe*. Retrieved January 31, 2022, from https://ecfr.eu/article/turkeys-drone-diplomacy-lessons-for-europe/

Gambhir, M. (2021). *Defense Diplomacy and its Relevance*. Retrieved July 28, 2022, from https://www.claws.in/defense-diplomacy-and-its-relevance/

Grattan, R. F. (2011). *Strategic Review: The Process of Strategy Formulation in Complex Organisations*. Gower Publishing.

Grina, G. (2017). National Military Diplomacy and its Prospects. *Lithuanian Annual Strategic Review*, *15*(1), 153–177. doi:10.1515/lasr-2017-0007

Haber, T. R. T. (2022). *Türk teknoloji şirketinin askeri metaverse uygulaması ilk kez NATO'da tanıtıldı*. Retrieved August 14, 2022, from https://www.trthaber.com/haber/bilim-teknoloji/turk-teknoloji-sirketinin-askeri-metaverse-uygulamasi-ilk-kez-natoda-tanitildi-701333.html

House of Commons. (1998). *The Strategic Defense Review White Paper*. Retrieved 28.07.2022 from https://researchbriefings.files.parliament.uk/documents/RP98-91/RP98-91.pdf

Khern, N. C. (2009). On command. *Pointer: Journal of the Singapore Armed Forces Supplement*, *34*(1), 1–28.

Koerner, W. (2006). Security sector reform: Defense diplomacy. *Library of Parliament. Parliamentary Information and Research Services Brief*, *6*(12), 1–3.

Lamsal, H. L. (2022). Effectiveness of military diplomacy towards nationalism, national security, and unity. *Unity Journal*, *3*(01), 82–96. doi:10.3126/unityj.v3i01.43317

Lemon, E., & Jardine, B. (2021). Central Asia's multi-vector diplomacy. *Kennan Cable*, (68), 1–12.

Mearsheimer, J. (1990). Back to the future: Instability in Europe after the cold war. *International Security*, *15*(1), 5–56. doi:10.2307/2538981

Metin, M. (2022). *Turkish defense industry gains its interdependence.* Retrieved August 16, 2022, from https://businessdiplomacy.net/turkish-defense-industry-gains-its-independence/

Midhio, I. W., & Priyono, J. (2019). Education and research as components of Indonesia's defense diplomacy. *Jurnal Pertahanan*, *5*(1), 61–70. doi:10.33172/jp.v5i1.487

Miller, D. C. (2015). *SIMNET and Beyond: A History of the Development of Distributed Simulation.* Retrieved July 29, 2022, from https://www.iitsec.org/-/media/sites/iitsec/link-attachments/iitsec-fellows/2015_fellowpaper_miller.ashx

Milli Savunma Bakanlığı. (2021). *2020 yılı faaliyet raporu.* Bütçe ve Mali Hizmetler Genel Müdürlüğü.

Milli Savunma Bakanlığı. (2022). *Yurtdışındaki askeri ateşelikler.* Retrieved August 15, 2022, from https://www.msb.gov.tr/SavunmaGuvenlik/icerik/yurt-disindaki-asker-ataselikler

Muniruzzaman, A. N. M. (2020). Defense diplomacy: A powerful tool of statecraft. *CLAWS Journal*, *13*(2), 63–80.

Nathan, J. (1993). Force, order, and diplomacy in the age of Louis XIV. *The Virginia Quarterly Review*, *69*(4), 633–649.

NATO. (2022). *Extraordinary virtual summit of NATO Heads of State and Government.* Retrieved July 29, 2022, from https://www.nato.int/cps/en/natohq/news_192453.htm

Nato. (2020). *Partnership for Peace programme*. Retrieved July 28, 2022, from https://www.nato.int/cps/en/natohq/topics_50349.htm#:~:text=The%20 PfP%20was%20established%20in,every%20field%20of%20NATO%20 activity

Nicolson, H. (1941). *Diplomacy*. Oxford University Press.

Nye, J. (2005). *Soft power: The means to success in the World politics*. Public Affairs.

Plessis, A. (2008). Defense diplomacy: Conceptual and practical dimensions with specific reference to South Africa. *Strategic Review for Southern Africa, 30*(2), 87–119.

Report, A. W. (2021). *Drone technology propels Turkey's defense diplomacy and exports*. Retrieved December 9, 2021, from https://aviationweek.com/ aerospace/emerging-technologies/drone-technology-propels-turkeys-defense-diplomacy-exports

Resdal. (1994). *Germany White Paper 1994*. Retrieved July 27, 2022, from https://www.resdal.org/Archivo/d0000066.htm

Singh, J. (2021). Military diplomacy: An appraisal in the Indian context. *CLAWS Journal, 15*(2), 108–124.

Sputnik. (2022). *ABD ile G. Kore, online askeri tatbikatla damarına bastıkları K. Kore'ye 'masaya dön' çağrısı yaptı*. Retrieved July 29, 2022, from https:// tr.sputniknews.com/20220418/abd-ile-g-kore-dijital-askeri-tatbikatla-damarina-bastiklari-k-koreye-masaya-don-cagrisi-yapti-1055656984.html

Thinktech, S. T. M. (2022). *Yeni bir paradigm: Askeri metaverse ve geleceği*. Retrieved July 29, 2022, from https://thinktech.stm.com.tr/tr/yeni-bir-paradigma-askeri-metaverse-ve-gelecegi

Thorpe, J. (2010). *Trends in modeling, simulating & gaming: Personel observations about the past thirty years and speculation about the next ten*. Retrieved July 29, 2022, from https://www.iitsec.org/-/media/sites/iitsec/ link-attachments/iitsec-fellows/2010fellows_thorpe.ashx?la=en

Türkiye Cumhuriyeti Dışişleri Bakanlığı. (2022). *Türkiye Cumhuriyeti Dışişleri bakanlığı tarihçesi*. Retrieved August 15, 2022, from https://www.mfa.gov.tr/turkiye-cumhuriyeti-disisleri-bakanligi-tarihcesi.tr.mfa#:~:text=1924%20y%C4%B1l%C4%B1nda%2039%20 d%C4%B1%C5%9F%20temsilcili%C4%9Fe,toplam%20253%20misyona%20 sahip%20bulunmaktad%C4%B1r

White, H. (2014). Grand expectations, little promise. In B. Taylor, J. Blaxland, H. White, N. Bisley, P. Leahy, & S. S. Tan (Eds.), *Defense Diplomacy Is the game worth the candle?* (pp. 10–11). ANU Strategic and Defense Studies Centre.

Winger, G. (2014). The velvet gauntlet: A theory of defense diplomacy. In What Do Ideas Do IWM Junior Visiting Fellows' Conferences (vol. 33, pp. 1-15). Vienna: IWM.

Yackley, A. J. (2020). *How Turkey militarized its foreign policy*. Retrieved October 15, 2020, from https://www.politico.eu/article/how-turkey-militarized-foreign-policy-azerbaijan-diplomacy/

KEY TERMS AND DEFINITIONS

Diplomacy: The art of conducting International Relations. It has a long history.

Digitalization: The process of transferring accessible information to digital media that can be read by any computer.

Public Diplomacy: A process that describes how the government of one country tries to influence the citizens and intellectuals of another country in line with their own political and ideological ideas.

Digital Diplomacy: Method of solving foreign policy problems via the internet

Defense Diplomacy: Achieve foreign policy objectives through the peaceful use of the state's military and security apparatus

Military Diplomacy: Although it is used instead of defense diplomacy, it is a sub-title of defense diplomacy.

Armed Unmanned Aerial Vehicle: A type of unmanned aerial vehicle that carries aircraft munitions such as bombs, missiles and/or ATGMs and is used for drone attacks.

Unmanned Aerial Vehicle: A type of flying vehicle that is physically unmanned

Chapter 5

China's Coronavirus-Oriented Diplomacy in Nigeria:
A Content Analysis of the Chinese Embassy's Online Communication

Chinonso Aniagu
Independent Researcher, Nigeria

ABSTRACT

Over the last decade, new media penetrations have multiplied the determinants of perceptions held in diplomatic sphere. The deployment of digital diplomacy by different countries to manage public perceptions has now opened a new vista in diplomatic and public relations discourse. China's application of digital diplomacy became even more pronounced as a result of the negative narratives--mostly derived from preconceived stereotypes--that heralded the outbreak of coronavirus. Consequently, the country's online communications in Nigeria were designed to launder its image for a more beneficial bilateral relationship with the latter. This chapter, through qualitative content analysis, evaluates the extent China's online coronavirus-oriented communications in Nigeria were deployed in countering negative stereotypes and narratives at the initial-to-peak stage of the virus. Findings show that, despite trickles of counterproductive content, post-coronavirus China's image was significantly laundered through the deployment of various Nigeria-targeted digital communication strategies.

DOI: 10.4018/978-1-7998-8394-4.ch005

INTRODUCTION

The last quarter of 2019 unraveled a swift shift in the battle of the world super powers to diplomatic war buoyed by the resultant effects of coronavirus outbreak. The virus was deployed as an efficient content to score political points, de-market competitors and win new allies. Unsurprisingly, decade-long rivals, the United State of America and China, switched to their most potent communication diplomacy machinery to make the most of the situation and gain at each other's expense. The Asian giant, however, found itself at the wrong end of the diplomatic war and ensuing propaganda as it was credited with being responsible for the outbreak of the pandemic. The then president of the United States, Donald Trump took the Chinese "indictment" further by branding the virus "China Virus". Not wanting to be outdone in the communication and media circle, China activated its own communication strategy which was geared towards micro-managing the information and news coming out of the country, playing down the negative contents and countering the US-propagated narratives on the pandemic. The country, in a seeming effort to effectively implement this strategy, energized its diplomatic missions to deliver and disseminate strategic traditional and new media communication contents geared towards image laundering in some strategic- and most especially, developing- countries of interest around the world. Nigeria, with huge potential for bilateral socio-economic engagements, falls into this category of China's "countries of interest."

Evidence abounds that the White House views China's economic rise and increasing global engagement in countries like Nigeria as a danger to its global economic and political dominance. Suffice it to say that notwithstanding the fact that the idea of an unstoppable Chinese economic, military expansion and a relative loss of power for the United States could be said to be based on fairly tested assumptions and projections, China seems to, genuinely, be the only country with the potential to threaten the status of the United States (Perthes, 2010). It is not, therefore surprising that the US is doing everything to whittle down the former's growing global influence- an influence touted be becoming more pronounced in Africa which is widely considered a huge potential in the scheme of world economy. Underlining the fact that Africa is at the center of US-China rivalry Mulualem (2013) noted that the US and China boost of the first and the second largest economies in the world and, though neither intimate friends nor fierce enemies, have developed foreign policies targeted towards capturing Africa.

Though the battle to "win" Africa has become fierce lately, the two countries are said to have exercised decades of "soft power" towards the continent having engaged in volunteer activities over the last decade. In what seems like a decisive step to curb the US influence on the continent, China, in 2004, launched Overseas Youth Volunteer Program- a development that looked a deliberate balancing measure for the US Peace Corps which has been operating in different African countries since 1960.

Despite the strife by China usurp US dominance of the African market, the former remained firmly rooted in the continent heading to the end of the nineteen century owing largely to the fact that its cultural influence was more pronounced than that of the latter in most part of the continent. The impact of establishments like Hollywood and CNN in African countries like Nigeria has been enormous as much as hundreds of sister city relationships between the country and African cities and towns (Mulualem, 2013).

With China scaling up its commercial and subtle political engagements in the Africa, the country has now become most important partner of the continent (Taylor, 2007). While these engagements have given China significant foothold in African countries like Nigeria, issues associated with such enterprises have been more or less problematic and potentially destabilizing. Chinese businesses in the country are widely associated with negative portrayal of substandard product manufacturing while activities such as solid minerals mining have been blamed for exacerbating home-grown conflicts, mis-governance, and policing failures. (Page, 2018). It has, however, been difficult gauging the facts of this perception with such views said to be enabled and propagated by the US and Western world who seem to already have biased outlook of the China's global business policies.

China, having identified itself as a target of Western de-marketing strategies, has, therefore become increasingly desperate to shore up its brand and present a very good image as the country extends its global footprint (Tse & Hung, 2014). The country has continued to invest heavily on selling positive image of its associations with other countries, not only to counter the ani-Chinese, pro-Western narratives but to ensure that its foreign policy objectives are met. With culture being a key ingredient of both nation branding and public diplomacy, the country, following the examples of British and French investments in British Council and Alliance Francaise, established China's Confucius Institutes to engender effective brand management and loyalty (Copeland, 2012).

Adding to the rapid expansion of Confucius Institute by nearly 400 branches since 2004, other considerable developments that have been carried out over the past decade to promote the country as an attractive and trusted member of the international community the contribution of more than 3,000 troops to serve in U.N. peacekeeping operations, participation in multilateral talks, and the hosting of mega events, including the Beijing Olympics and Shanghai Expo in 2008 and 2010, respectively (Hung, 2015)

While the use of other aspects of digital communication by the Chinese Government to launder its image has yielded positive result, it is the deployment of digital diplomacy that has massively been relied on to market the country and ensures that negative reportage about and against it is "neutralized". The postmodernist revolution in diplomacy, having engendered fundamental changes at an unprecedented rate- that which affected the very character diplomacy have always been known for ("German Institute", 2018), has affected every aspect of China's international political engagement.

The country's foreign ministry and diplomatic team have now adopted new media-oriented channels to ensure seamless operations and effective delivery of communication to strategic targets. The tilt towards digital diplomacy is understandable in that it not only blends with bilateral connection, which is now becoming popular preference over the multilateral approach that was heavily relied on by many countries in the 90s, but provides a potent platform for engaging directly with the citizenry. With the direct public involvement of the populace—often mediated by social media— placing demands on diplomatic-related contemporary agenda becomes easier to achieve ("German Institute", 2018).

Consequently, this chapter, deriving its theoretical base from the Image Repair/Restoration and Framing Theories, aims at analyzing the China's government online communication targeted at "valeting" the country's image in Nigeria at the start of the pandemic (February to August, 2020). The chapter will, in a nutshell, qualitatively content-analyze China's deployment of communication diplomacy through a tripartite lens to explore the negative stereotypes of China that emerge as a result of the outbreak of the COVID-19 pandemic in Nigeria, show how through social media communication, the Chinese Embassy sought to launder its image in Nigeria and examine the extent to which China's effort were effective.

There has also been various academic research on some aspects of China's coronavirus diplomacy in the recent past. While Gauttam, Singh and Kaur (2020) focused on China's deployment of its global health diplomacy as a means of wielding soft power, Liu, Huang and Jin (2022) concentrated on the effect the Sino-American geopolitical competition is having on China's COVID-19 vaccine diplomacy. Lang (2019) provided a more strategic perspective balancing of China's integrity outlook against existing stereotypical evaluations. Madrid-Morales (2017) was, however, closer to the discourses of this chapter with a more Afrocentric evaluation of China's digital public diplomacy. This is, however, novel, in that it weighs China's coronavirus diplomatic communication against specific Nigerian-oriented stereotypical context.

METHODOLOGY

The samples for the content analysis are all publications made on the Embassy of the People's Republic of China in the Federal Republic of Nigeria website, http://ng.china-embassy.gov.cn/eng/zngx/cne/index_1.htm. These online publications- twenty-five (25) in number- were uploaded to the website between February and August, 2020 and also shared across the embassy's social media pages. While the number of sample was determined by availability within specified scope, the period covered fall within what was considered initial-to-peak period of the pandemic. The content analysis is qualitative and interrogated the deployment of digital communication contents to launder China's image to counter the prevalent negative perceptions that are driven by a blend existing stereotypes and COVID-19 narratives. The effectiveness of these of communication and image laundering strategies deployed by the Chinese Embassy will be weighed using documented resultant reaction of Nigerian government officials and other state actors.

THEORETICAL FRAMEWORK

Image Repair Theory

The importance of having good image in the comity of nations cannot be overemphasized. The 21st century bi- and multi-lateral engagements have seen countries all over the world explore different revolutionary communications mechanism to not only create good brands but also react to contents that are capable of derailing the positive optics of their nations. China- a country with huge growth ambition- have had to device a brand management and communication crisis approach that, over the years, have been more reactionary to launder its image against prevailing brand crisis. In analyzing China's image laundering mechanism, the image repair theory will provide a appropriate theoretical base, owing to its deep correlation with the tenets of restoring a depreciating brand.

The theory- also referred to as restoration theory states that a bad image can be improved but not necessarily restored in totality (Benoit, 2015). This chapter will evaluate China's COVID-19 digital diplomacy strategies and deployment of the theories categories like denial, evading responsibility, reduction of offensiveness as well as correction and mortification. The usage of these broad categories will be weighed against negative stereotypes which mainly borders on anti-democratic postures and pro-profiteering attitude- even in the alter of human dignity- considered offensive by humanity.

Chinese Embassy's online communications to Nigeria are also have the fundamentals of the image theory embedded in them as their goals are centered on changing the negative perceptions to not only the Nigerian government but the citizenry that has become prevalent in the global space as a result of its connection to the pandemic. The huge investment on laundering China's image within the period under review goes to show that the country understands that reputation, when not compromised, does the fundamental job of engendering further national progress.

Framing Theory

While the image repair theory is centered on wider confines of brand communication management, the framing theory is more channel-specific and focuses on the deployment of media to transport preferred information to the public. The theory posits that the interpretation of certain information is largely dependent on media packaging and highlights of event that created the narrative of the content. The proponent of the theory is of the view that the media have the power of influencing how the contents are interpreted at the decoding stage. China's choice of online platforms for diplomatic interactions with Nigeria allows the former to not only control the information being accessed but also ensure that such contents are deployed towards laundering its image. It is not a coincidence; therefore, that China's online platform chose to flaunt its positive economic contributions to Nigeria at the peak of COVID-19 when it seemed as if its image was being overwhelmed but that dark edge of the pandemic.

The online platform also created frames that were aimed at influencing Nigeria and Nigerians to conjure the image of a "selfless partner" once "China" is mentioned. This was done through multiple emphases on what the Asian giant was doing to save the rest of the world from feeling the strain of the pandemic. Several repetitions of China's investments in Africa were also used to bring to the fore the humanitarian efforts by the country to help the continent out of past health crises.

ANTI-CHINA CONSPIRACY THEORIES AND STEREOTYPES

China and Negative Stereotypes

That China found itself at the wrong end of the COVID-19 controversies and resultant conspiracy theories was not down to immediate factors but largely because of long-perceived stereotypes which the circumstances surrounding the outbreak and spread of the virus only affirmed. These factors, when broadly delineated are stereotypical re-presentation as inhumane and pro-profiteering nation. While lesser factors like racism and disrespect for democratic tenets are grouped under the inhumane stereotype, counterfeiting and materialism are captured under profiteering.

Political leanings, in the 21st century, have become such a deciding factor in information encoding and decoding. Conspiratorial endorsements and acceptance of misinformation are generally believed to be an offshoot of political leanings. Dissemination of fake news during the 2016 US presidential election was, for instance attributed to political conservatism of citizens of 65 years and above which points to potential effect of low level of digital literacy (Agley & Xiao, 2021). A lot is put into ensuring that media texts and information are re-presented to sell a particular ideology or tilt towards a particular bent. At the decoding stage of such information too, consumers are now emotionally equipped to decode in line with innate ideologies. These ideologies and worldview are heavily influenced by existing stereotypes. These stereotypical representations in the subconscious are the base of a decoders "fact." Long before COVID-19, China had already been re-presented as a country which have no respect for human dignity and whose foreign policy is built on selfish, one-sided profiteering that is only meant to exploit the less developed countries. These stereotypes have been illuminated and reechoed by the Western media who have consistently cited the limitations and inability of the China's pro-communist regime to fall within the confines of "universal" democratic tenets as a poster for the anti-human world the latter seems pumped to engender.

Furthermore, the negative memories of colonialism have awakened the consciousness of Africans to guard against similar occurrence in the future. The colonial administrators were marked with the notoriety of crude exploiters who took advantage of the continent, culminating in the very negative history of slave trade. The diplomatic inroads made by China in Africa are viewed with maximum suspicion as the continent strife to avoid re-colonization. Africans, therefore, view prospective foreign partners from racial lens- most of the time borne out of the quest to demonstrate strength and draw even with races hitherto considered superior.

Despite the high level of awareness to guard against racial exploitation, China's Africa policy is said to be characterized by classic realism in that the country, just like other Western nations, are attracted to the continent as a result of its huge economic opportunities and markets. Though some evident concessions are made at national stage to accommodate the culture of these African countries, racial inequality becomes visible at the individual and corporate levels (Shih, 2013).

The perception of China as a country where racial discrimination thrives was that which gained traction over time. The anti-black sentiment by Chinese student which started gathering momentum in 1970s culminated the anti-African students Nanjing protests in 1988 and 1989. The protest, which received wide media coverage, was widely seen as xenophobic in view of the fact that the small African community was considered not part of the mainstream Chinese social order. The protest was a brutal reminder of ethnic conflicts that plague the larger part of the globe (Sautman, 2009).

China may have inadvertently built foundation for such perception owing to its inability to allow for inclusivity in its management of the fifty-six ethnic groups. Chinese are perceived as a country which sees any other culture or race different from theirs as inferior, laying credence to the nationally held opinion that they are the most civilized people in the world

Despite the swelling numbers of other ethnic groups in China today, the Han Chinese have been imposed as the leading race in the country, ensuring that the "Han" identity is uncompromised in the country. The application of eugenic policies allows a forceful biological implementation of this single race entity with actions as extreme as abortion taken to ensure that there is no "adulteration" of Chinese blood (Dikotter, 1992). This not only goes against the right of individual to make telling choices but bolsters the perception of China as a country that champions racial discrimination.

It is, therefore, unsurprising that China was accused of meting out racial hostility towards Africans/Blacks in the southern Chinese city of Guangzhou while the world was battling the second wave of the pandemic in April 2020 (Zhu, 2020). It is ironic that Guangzhou, a city with largely undocumented African community of over 200,000- popular among African immigrant seeking greener pastures (BBC News, 2020; Lan, 2016)- gained negative publicity at such a time China was trying to manage its coronavirus-battered image. The widely reported incidents in Guangzhou became a metaphor of the anti-black racism in China that has been hidden from the rest of the world before now; that which runs counter to the country's official self-adulation about respect and support for Africa(ns) (Tettey, 2021).

The ill-treatment of Africans in Guangzhou which was described vulgar racism, and exploitative only reaffirmed China's pro-racial discriminatory status. The Black community in the country was, according to Quassini and Quassini (2022), all the while reported to have been targeted with multiple racial projects by Chines authorities, labeling their bodies as diseased and physical presence as a threat to the viability and safety of the Han who are in the majority. With Nigerian nationals being at the center of these reports, the country's Ministry of Foreign Affairs had to react in a dissenting manner to the global outcry which saw citizen journalist mobilize an anti-China movement to expose the discriminatory actions taken against the Black community in country.

The re-presentation of China as an inhumane nation also influenced the perception of tge country as that which considers excessively profitable trade and business ventures over humanity. The profiteering stereotypical perception of China in Nigeria is largely down to the large patronage of Western media by the citizenry and the magnification of the former's negative realities through mind- penetrating re-presentation. China has been at the centre of the global menace of counterfeiting with the country being largely held culpable for the greater percentage of crimes associated with same.

Efforts by the central authority in China to fight counterfeiting are said to being drowned by the institutionalisation of the crime by local authorities. According to Andrew C. Mertha (2005), Beijing's good intentions have, in the past fallen short of the local governments' activities who are often the beneficiary of IPR violations. It is, however, not wrong that this stereotype may have been formed largely on the corridor of perception more than verifiable evidence as Chow (2000) maintained that there is no scientific formula of determining the quantum of counterfeiting activities taking place in China. Devising a means to counter this perception, China acceded to WTO's Trade Related Aspects of Intellectual Property Rights (TRIPS) Agreement, an action Fleming attributed to being largely image-related (2020). This has, however, done little to change to Chinese perception in Nigeria and Okafor (2020) asserted that most of the fake products being imported into the West African country were being shipped in by China.

Furthermore, while interrogating China's interest in Nigeria, a 2018 Special Report by the United States Institute of Peace stated that China's relationship with Nigeria does not fit into the paradigms of mutually beneficial collaboration but a strategic interest to profit from Nigeria's status as an emerging global power, being the largest economy in Africa, a major oil and gas producer, and prospective world's third most populous country by 2050 (Page, 2018). Creating a metaphor of how close the relationship between Beijing and Africa has become, the report notes that top officials of Chinese Government rarely miss an opportunity to stop off in Nigeria when visiting Africa- a practice their Nigerian counterpart now seem to be reciprocating by the constant visits to Beijing.

China is, therefore, predominantly seen by Nigerians as a country who will do everything to make profit even when it involve dishonestly undertaking a sharp practices that include but not limited to anti international trade actions like counterfeiting. This stereotype gave foundation to the approaches deployed by the citizenry in the decoding and interpretation of China's COVID-19 news and communication. In other words, the pre-pandemic biases against China played a major role in acceptance or rejection of contents, theories and narratives thrown up by the outbreak of the virus.

Coronavirus Origin and Anti- China Conspiracy Theories

Establishing the widely publicized information that the virus originated from China, WHO stated that all the published genetic sequences of SARS-CoV-2 isolated from human cases are "very similar, suggesting that the start of the outbreak resulted from a single point introduction in the human population around the time that the virus was first reported in humans in Wuhan..." (2020, p.1). While WHO's definiteness on the part of the world the first case of the virus was confirmed did little damage to the China's reputation, its publication the genetic sequences analyses which confirmed that the transfer from an animal source to humans occurred in Q3 of 2019 fueled anti-China conspiracy theory.

As soon as the first cases of COVID-19 were reported in late December 2019, investigations were conducted to understand the epidemiology of COVID-19 and the original source of the outbreak. The report specifically stated that the investigation conducted to understand the epidemiology of the virus immediately the first cases were recorded confirmed that a significant proportion of the initial cases (between December 2019 and January 2020) were directly linked to the Huanan Wholesale Seafood Market in Wuhan City- a market with very high patronage of seafood, wild, and farmed animal species.

In an accusatory tone that seem to amplify the widely held belief on China's complacency, WHO was categorical in pointing to that fact that environmental samples taken from the Wuhan City market in December 2019 confirmed that the city played a major role helping the virus gain momentum at the initial stage of the pandemic.

The World Health Organisation (2020, p.1), in a subsequent official account of the origin of virus, traced the virus to SARS-CoV-2 isolated from humans which have been proven to be closely related genetically to coronaviruses isolated from bat populations like bats from the genus Rhinolophus. It also traced the virus to the original 2003 SARS outbreak, which is also closely linked to coronaviruses isolated from bats and whose genetic relations suggest related ecological origin with bat populations. Agreeing to the WHO account, Alanagreh et al. submitted that the virus was similarly named SARS-CoV-2 as it belongs to the Betacoronavirus group- same group with the 2003 severe acute respiratory syndrome coronavirus, SARS-CoV (2020).

In what seems like an image laundering exercise to pacify China who is its major financial contributor, however, WHO went ahead to emphasize the universal existence of Bats in the Rhinolophus genus by stating that they are not only found in Asia but also across Africa, the Middle East, and Europe. To diffuse the theories that the virus may have come from Chinese association with domestic animals, WHO insisted that SARS-CoV-2 is not genetically related to other known coronaviruses found in farmed or domestic animals. It goes ahead to state that investigations into the first human cases of the virus have determined that carriers they had onset of symptoms as at December 1, 2019 but refused to directly link them to the Huanan Wholesale Seafood Market, noting that they may have contacted the virus in November through contact with earlier undetected cases (incubation time between date of exposure and date of symptom onset can be up to 14 days). Clearly exonerating China of deliberately introducing the virus as were being adduced by many conspiracy theories, WHO further explained that it is not possible to determine precisely how human beings first contacted the virus in China but asserted that "all available evidence suggests that SARS-CoV-2 has a natural animal origin and is not a manipulated or constructed virus".

To underline the magnitude of the effect of these conspiracy theories, the WHO Director General, Tedros Ghebreyesus, early 2020, cautioned that the world had gone from fighting epidemic to waging a war against infodemic, noting that fake news spreads faster and easily more than coronavirus (WHO, 2020). Similarly, a United States survey conducted in March 2020 reported that 42% US citizens have been availed of news about the virus that is completely fabricated (Mitchell & Oliphant, 2020).

The fabrications of the different conspiracy theories for and against China are down to a lot of factors ranging from information void and existing stereotype. Being a novel virus, it is not surprising that, in the absence of specific information as to what its origin was, the world was in to a long ride of myriad of conspiracy theories created to fill the void temporary opened by inability of responsible authorities to provide authentic information.

The uncertainty about the precise origin of the virus, adding to the natural inclination of human to apportion blames, no doubt, contributed to the creation of numerous conspiracy theories which surfaced at the initial stage of the pandemic (Van Bavel et al., 2020). One of the notable theories that were formulated as a result of void created by lack of clarity in the origin of the virus and bought into by Nigerians within this period includes that which claimed influential societies invented the pandemic so as to take advantage of it to impose tracking technology which will be inserted as chip- unconsented to- in the masses during vaccination (Burnard & Richards, 2020). Another theory that equally made waves within this period was the Event 201 Theory which states that an event organized by John Hopkins Centre for Health Security in October 2019- and attended by various stakeholders or players that included representatives of the health sector, the private business sector, politicians, governments, Centers for Disease Control and Prevention and the World Health Organisation- was where the real pandemic was "planned" (Pearce, 2020). While these conspiracy theories seem anti-West when deconstructed, a lot of others were propounded to elicit anti-China emotions when decoded.

One of the unproven anti-China theories the absence of authoritative information on the origin of the virus gave room for at the start of the pandemic is conspiracy theory about coronavirus being a biological virus manufactured from a Chinese laboratory. While some disseminators of these theory posited that such virus was a result of a biological experiment that went wrong, some others accused China of manufacturing and unleashing the virus on the rest of the world to cause mayhem, instill fear and force the world to succumb to patronizing the vaccine market it was planning to create. This account, which specifically stated that the virus was created in a Wuhan laboratory, has not been substantiated and at variant with study such Bolsen and Palm's "Framing the Origins of Covid-19" (2020) which reported that research respondents beliefs that China was responsible for the virus are tied to their "willingness" to penalize the latter for the culpability of its government in the creation of the pandemic.

Similarly, Vincent (2020) noted that many societies witnessed viral phenomena of fake news- some of the most prevailing being that which attributed the cause of the virus to 5G cellular technology- at the same rapid and aggressive speed at which the virus was spreading across the globe. The 5G-based conspiracy theory was constructed around the narrative that transmitters from the network, through radiation, make people to fall sick to the extent that such sicknesses lead to their eventual deaths. While this theory lacked any scientific proof, it is instinctive to note that it was "adapted" by some major public opinion molders in Nigeria. The fact that the negatives of the network were attributed to China, who only introduced the network to its shores after it was first deployed by South Korea, goes to show that the Chinese image was the primary target of the proponents of the theory.

COUNTERING THE BLEND OF ANTI-CHINA COVID-19 NARRATIVES AND NEGATIVE IN NIGERIA

Countering Anti-China COVID-19 Origin and Conspiracy Theories

To counter the conspiracy theories that held China culpable for the pandemic, one of the reactions of the Chinese Embassy in Nigeria was captured in the "Opening Remarks of Ambassador Zhou Pingjian of China at the Embassy's Media Briefing on Fighting 2019 n-CoV". The press statement was the first COVID-19-related publication on www.ng.china-embassy.gov.cn/eng and comprehensively dealt with issues concerning the virus through China's lens.

An analyses of this publication shows that the Chinese Ambassador to Nigeria, Zhou Pingjian, centred his communication in the document around contents encoded as brand crisis mechanism to counter the widely publicized narratives on the origin and the negative role played by his country leading to the spread of the virus. Pingjian emphasis that the fight against the virus was of a common concern of the international community could be seen as an attempt to de-emphasis the negative concentration on China at the initial stage of the pandemic. To puncture the already-publicised anti-China account on virus' origin, the Ambassador directly quoted a portion of the WHO (2020) document where the DG, Tedros warned the media to avoid unverifiable narratives about the pandemic as it was a "time for facts, not fear… time for science, not rumours…time for solidarity, not stigma". This particular quote could be seen as a deliberate act to set the tone for counter narrative that is meant to launder China's image amid series of damaging publicities that followed the outbreak of the virus.

The said damaging publicities were mainly created to answer questions regarding the origin of the virus which, as at that time, was shrouded in a lot of ambiguities and secrecy. In the bid to clear some of these ambiguities that surrounded China's activities and culpability leading up to the outbreak of the virus, Pingjian, in February, 2020, published another article entitled "NCP: Time for Facts, Science and Solidarity" on the Embassy website to further advance a pro-China narrative and counter the anti-Chinese theories that have been situated at the centre of global discussions. The article, first published in leading Nigerian newspapers, Thisday, Leadership, New Telegraph, Guardian, as well as Sun, People's Daily and The Nation was a more elaborate attempt to launder China's image and strategically manage the negative narrative that were already beginning to mask the latter's integrity. Pingjian unequivocally countered what China considered "horrible" rumour about the virus meant to create global panic, with Nigeria at the centre of the adverse effect of such atmosphere. According to Pingjian (2020a),

Virus is horrible. What's more horrible is rumors and panic. As WHO Director-General Dr. Tedros has repeatedly called on people not to believe in rumors or spread them, "This is the time for facts, not fear. This is the time for science, not rumors. This is the time for solidarity, not stigma." In some countries including Nigeria, relevant departments have been stepping up efforts to bring to justice those creating and spreading rumors about the epidemic. On February 1, five Abuja residents who had faked coronavirus infection were swiftly arrested by the FCT police for their gratuitous prank. (para. 7)

It is obvious that the observation the unfavourable social media posts which was then feeding off misinformation and fake news, motivated China to device a more direct approach to discourage Nigerian citizens from partaking in what has become a popular trend. While the reference to the actions taken against purveyors of coronavirus fake news in Nigeria seems to be altruistic communication dissemination, a further deconstruction of the articles shows that this may have been born out of strife by China to effectively stifle the anti-Chinese narratives that were becoming dominant sentiment in the Nigeria.

Image Repair's denial strategy was also used in the article to deny that China was totally responsible for the outbreak of the virus. The publication used the WHO's inability to pin the virus on specific origin as an alibi to advance its argument that China could not have been found culpable of any activities leading to the global health crises. It particularly cited the WHO's acceptance that the animals that are being suspected to have transmitted the virus to humans are not only existing in China but in other parts of the world and continent.

Similarly, another article published on the Embassy website entitled, "Ambassador Zhou Pingjian: China's Response on COVID-19 Responsible, Decisive, Effective" illumined the fact that such virus are not only peculiar to China, citing past global health crises to prove that it is far less deadly than what has been witnessed in other parts of the world. The article stated that in contrast to the low mortality rate of coronavirus, the H1N1 flu of 2009 and the Middle East Respiratory Syndrome (MERS) of 2012 had mortality rate of 17.4% and 34.4% respectively (Pingjian, 2020b).

The most significant deployment of strategic image laundering communication in this were the reference of the 4% mortality rate of the Lassa fever which was prevalent in Nigeria and data of flu patients from the United States CDC. While article cited no less a Nigerian government official than the Minister of Health, Osagie Ehanire, to prove that the mortality rate of Lassa fever was about 30% a couple of years ago in the West African country, it emphasised that,

according to a recent United States CDC estimation, from October 1, 2019, through February1, 2020, in the US there had been 22-31 million flu illnesses, 10-15 million flu medical visits, 210-370 thousand flu hospitalizations, and at least 12-30 thousand flu deaths. (Pingjian, 2020b, para. 8)

The reference to the credible US institution here could be seen as a strategic attempt play down the effect of the virus and to lay credence to the fact that the latter is also susceptible to worse health crises. This is significant as China sees America as the country that is responsible for the overwhelming negative narratives it has received since the outbreak of the virus. Here, the Chinese Embassy employs Reducing Offensiveness strategy of the Image Repair Theory by reducing the degree of the pandemic so as to make it sound less scaring. The strategy is also in line with framing theory as the embassy carefully chose the content to to discuss in order to downplay the dominant negative narratives.

In a more detailed to online publication against the pro-US narratives on the outbreak of the virus, Pingjian (2020c) in his "Working Together towards a Health Silk Road" took a more direct communication approach to respond to the growing resultant fledging anti-China sentiment. He accused the US media of having ulterior motives by "falsely" claiming that the virus originated in China. Pingjian article which was also published in local Nigerian newspapers was, no doubt, a diplomatic strategy to reverse what was becoming popular narrative of China's role in the outbreak of coronavirus. This perception was in a large scale aided by the opinion of the US President, Donald Trump, who demanded a formal apology from China and even went ahead to refer to the virus as Wuhan Coronavirus which according to Pingjian was against the WHO guideline.

Extending the face-saving, image laundering argument in the publication, Pingjian in a persuasive tone dug into the US-narratives advancing four rhetorical questions;

What is the point in arguing that someone should apologize for it? We see people around the world joining hands to fight off this epidemic, and the WHO has repeatedly said that stigmatization is more dangerous than the virus itself? Why are certain people and media still promoting such an absurd logic? What are they up to? (2020c, para. 11)

A further deconstruction of the questions asked shows that they were meant to evoke the feeling of guilt from Nigerians who have bought into the anti-China conspiracy theories without recourse to WHO's status which clearly stand against stigmatization. The questions also serves as instrument of solidifying the fact that there are some agenda being promoted by the media meant to achieve some predetermined end considering that health crisis such as H1N1 flu that broke out in the US in 2009 and killed at least 18,449 people in a year was not given the same colouration (Pingjian, 2020c).

The reference of the relatively low mortality rate of COVID-19 at this point was down to the deployment bolstering strategy of the Image Repair Theory. This was internationally done by the Chinese Embassy to increase the positive sentiment towards the virus by showing that it has killed far less number of people compared to other historical health crises with far higher mortality rate.

It is also evident that the main communication strategy adopted by China to launder its image in the face of the negative COVID-19 narrative was built around constant repetition and reiteration of what it considers its unique selling point. This was obviously reflected in most of the digital publication targeted at the Nigerian audience through the diplomatic website of the Chinese Embassy in Nigeria and other strategically selected new media platforms. Pingjian's, for instance, dedicated significant portion of his "Working Together towards a Health Silk Road" to showcase China's strength and highlighting its contributions to the international community. According to him,

China has conducted itself as a responsible country. China's signature strength, efficiency and speed in this fight have been widely acclaimed. To protect the health and safety of people across the world, the Chinese people have made huge sacrifice and major contributions. The international community commended China for its effective and extraordinary response and enormous sacrifice. (2020c, para.13)

Similarly, Pingian also stated that WHO's declaration of the outbreak as a public health emergency of international concern (PHEIC) was not as a result of China's inability to locally manage the health situation. Conveniently quoting WHO DG, Tedros, Pingjian argued that the main reason for such declaration was not because of happenings in China but a decision to protect the countries with "weaker health systems…which are ill-prepared to deal with potential spread" (Pingjian, 2020c para. 14).

A change of strategy was also observed at the later stage of the pandemic with the Chinese Embassy deciding to deploy a rebranding mechanism for Wuhan- a city, owing to President Trump's choice of words, became synonymous with COVID-19. Understanding that giving the city a new image is what is needed to change the perception of China in Nigeria, an article, "COVID-19: How CHINA Did It", was published online by the Chinese Embassy in Nigeria to paint the picture of recovery. The article highlighted what it considered the notable successes and turning point in the city's months-long battle with the virus, sending the message of hope to the rest of the world who are still grappling with the pandemic. As was the tradition in most of the online media texts published by the Chinese authorities within this period, the Embassy quoted the World Health Organization (WHO) press statement saying that "Wuhan provides hope for the rest of the world that even the most severe situation can be turned around." (Pingjian, 2020d, para. 4) Concluding, the said recovery in Wuhan was also likened to the situation in other parts of China were it claimed life and work has "quickly returned to normal" (Pingjian, 2020d).

It must be noted that there was no coincidence as to the time and objective of this publication. There is no doubt that this publication was targeted at countering the narrative that China was underreporting its new coronavirus cases. This was, ironically, at a time the numbers of new cases and death were increasing astronomically in Europe and the US. The publication was also targeted at quenching the doubts raised on efficacy of Chinese coronavirus pharmaceutical products.

Laundering China's Image against Negative Narrative and Stereotypes

Adding to the undemocratic, authoritarian image that China has been represented as in the media, the perception of country by Nigerians are heavily influenced by pro-counterfeiting and profiteering stereotype that it has been aligned with over the last decade. The phrase, "Made in China", among the Nigerian populace connotes counterfeiting and excessive pursuit of high profit margin. No doubt this stereotype played a major role in the choice of how communications and media texts were decoded and interpreted at the peak of COVID-19 pandemic. Chinese Embassy, well aware of this perception, set out to counter it and attempt to influence a new perception built on altruistic undertakings and endeavours.

While a natural disposition of the Chinese Embassy in Nigeria was to, through communication channels, counter this perception, an online publication from the Embassy's official website, "Opening Remarks of Ambassador Zhou Pingjian of China at the Embassy's Media Briefing on Fighting 2019 n-CoV" (Pingjian, 2020e), however, promoted same with part of the briefing playing down the magnitude of the virus by making reference to WHO insistence that, as at then, there was no pandemic yet or justification for measures that unnecessarily interfere with international travel and trade.

China's highlighting of the WHO opposition to the temporal shutting down of trade and movement only goes to affirm the widely attributed Chinese profiteering stereotypical perception which portrays the country as a country which puts businesses and profit-making over human lives. Suffice it to say that talking about trade in the alter of the impending global health crises in its first COVID-19 communication- meant to douse the already tensed global atmosphere- not only countered the objective the communication but became a subliminal re-assertion of the stereotype which China was trying to tune down within Nigerian opinion space.

Another instance where the Chinese authority inadvertently promoted this stereotype was in the joint statement released after the China-Africa Summit on Solidarity against COVID-19. The statement had highlighted "the importance of digitalization in the post-COVID-19 era and support efforts to speed up the development of Africa's digital economy and expand exchanges and cooperation on digitalization, information and communication technologies, especially tele-medicine, tele-education, 5G and big data" (FOCAC, 2020, para. 24). While the conscious effort to aid digitisation of Africa's communication sphere was, no doubt a pointer to China's intention to expand its horizontal in the continent through digital diplomacy, the attention given to trade in a communiqué drafted to update the continent on pro-Africa coronavirus interventions further twisted public opinions against the fore. More so, the inclusion of 5G network in the statement heightened the suspicion already created by the narrative of 5G conspiracy theorists.

In other words, the continued emphases on trade and economy, when looked at against the pre-existing stereotypical perception of a country driven by sole aim of excessive profit making, becomes counter-productive in that it inadvertently validates the predominant sentiment of the Nigerian government and citizens. The reference to 5G network also reawakened the consciousness of the populace to what was seen as script for implementation of all the projections of the 5G conspiracy theories.

In different layer of deconstruction, however, the constant reference to its trade dealings with Africa could be interpreted as a way to rechanneling the minds of the African governments to what it will benefit from China in terms material resources. This is an implementation of compensation strategy- another variation of the Image Repair Theory which holds that a guilty party, in order to restitution for crime committed, compensates the party who the said crime was committed against. Consequently, while China have been audacious in its denial that it was responsible for the outbreak of the coronavirus, it makes for a better strategy for it to subtle accept the significant portion of the blame and try to, even if indirectly, compensate African countries like Nigeria which are at the end of unquantifiable human and material losses.

In other publications on the embassy's website, however, China made effort to counter this stereotype by painting a picture of altruistic country who has continued to undertake selfless activities in Africa and the rest of the world. This was highlighted by the Pingjian (2020b) in his online publication "China's Response on COVID-19 Responsible, Decisive, Effective". The inclusion of "responsible" in the title of the article was meant to evoke pro-human imagery and portray the country as that which clearly pursues projects that promotes the well-being of humanity.

Pingjian, also used the article to emphasize that the health sector is a very important part of China-Africa cooperation, noting that 21,000 medical experts China sent to African in the past sacrificed their lives and contributed to the treatment of 220 million African patients. In a subtle reminder of its role during the outbreak of Ebola when the international community seemed to have deserted Africa, Pingjian states that,

After the Ebola epidemic raged through Africa in March, 2014, certain countries closed their embassies and evacuated diplomats and citizens from three West African countries hit by the epidemic. By sharp contrast, the Chinese government helped Africa at the earliest time possible. China sent not only urgently needed supplies but also medical teams of over 1,000 military and civilian doctors to areas stricken most severely by the epidemic. Chinese diplomats and medical experts chose to stay there instead of withdrawing. They fought together with local people until the virus was defeated. (2020b, para. 20)

Furthermore, China also mapped out online communication strategies to counter the narrative that China cared less about human. This explains the emphasis on "high importance" the country attach to lives and health of all people in China, Chinese and foreign nationals- as evident in "Opening Remarks of Ambassador Zhou Pingjian of China at the Embassy's Media Briefing on Fighting 2019 n-CoV". The mentioned publication also underlined China's acceptance in Africa, stating that Nigeria understands and has full confidence in its ability to control the virus. While this may not necessarily be statement of verifiable facts, it goes a long way to portray the West Africans as having taken side in the Chinese battle to sell its version of COVID-19 narrative against that being publicised by the Western media. It also serve as an indirect way of proven to its rivals that it has been able to sway the popular and public opinion in Nigeria to its favour despite the reported firm grips of the US-media on most of the African countries.

Similarly in the same publication, China, subliminally proving to Nigeria that it holds high value on human lives, stated the low infection rate of the virus around the world was as a result of the sacrifices it made to ensure that world survives the pandemic. According to the publication,

Guided by the vision of a community with a shared future for mankind, China is fulfilling its responsibility for the life and health of its own people and for global public health. Containing the epidemic at the source is of essence. China's effective response and sacrifice have averted the further spread of the virus in the world. The COVID-19 cases of infections worldwide is far less than 1% of the case count in China, while the H1N1 flu outbreak spread to 214 regions and countries. (Pingjian, 2020b, para. 13)

The Chinese communication strategies, which throughout the period reviewed have been characterised by a great degree of assertiveness and relative arrogance. Corrective action strategy was, however, adopted while dealing with issues of racism thrown up by the discriminatory treatment of Nigerianss in its city of Guangzhou. In a press statement released by the Press Secretary of the Embassy in reaction to the widely publicized incident, China states that

Ambassador Zhou Pingjian met with Rt.Hon. Femi Gbajabiamila, Speaker of the House of Representatives of Nigeria at his instance. During the meeting, the Ambassador was offered to watch a mobile phone video clip about the purported improper treatment on some Nigerian citizens in Guangzhou, China. The Ambassador made it clear that judging from the one video clip on that mobile phone the approach of the relevant epidemic prevention and control personnel was not inappropriate. We treat all foreign nationals equally in China. (Statement, 2020a, para. 1)

The adoption of corrective action strategy- as in Image Repair Theory- in its communication here could be said to be as a result of incontrovertible video evidence presented to its Ambassador by the Speaker of Nigerian House of Representative. It must be noted that this was one of the rare occasion where the Chinese Embassy accepted negative issue raised against it. It, however, strategically communicated its willingness to correct the situation and also gave assurance to return things to normalcy. While this proposition excites on the surface, one is moved to doubt the honesty of and in such promise as such gave the impression of a first-time happening. This is against the historical evidence that such discriminatory activities have been regular occurrence in the country

Just like criticism bordering on racial discrimination, issues transparency have been roundly used against China by the West to underscore the anti-democratic trait that it claims are predominant in government and culture of the communist government of the Asian country. The article, "NCP: Time for Facts, Science and Solidarity" by the Chinese Ambassador to Nigeria was also used to counter this narrative. According to the online publication, the "Chinese government has been sharing information in a timely manner, enhancing international cooperation, putting in place a nation-wide scheme, pooling national resources and taking the most strict and thorough measures to fight the outbreak" (Pingjian, 2020a, para. 6). The popular narrative in Nigeria had been that China, living up to the stereotype of been a non-transparent country, had some hidden agenda over the impending spread of coronavirus. To countered this, however, Pingjian (2020a) maintained that

China's openness, transparency and high sense of responsibility, and its decisive and effective measures have been recognized by the international community. Dr. Tedros Adhanom Ghebreyesus, the Director-General of the World Health Organization (WHO), affirms on many occasions that China has in fact taken more measures to contain the outbreak that it is required to do in case of an emergency, and China issetting a new standard for outbreak response for other countries. (para. 6)

The reference to the citing of the WHO's DG here could be seen as a strategic countering of the non-transparency narrative being advanced against China by international political rivals and bought by a large population of the countries bi- and multi-lateral partners. With WHO seen as a predominantly neutral and transparent body by the governments and people from these partner nations like Nigeria, it is only logical to think that it will be strategically more effective to build a favourable position of such "trusted" organisation while advancing an argument.

Similarly, Pingjian (2020f) in another article published in the Embassy's website and entitled, "COVID-19: Hardship Reveals True Friendship" cited the enhanced information disclosure by the Chinese government at the peak ok the pandemic to prove that the country has is transparent in his handling of the virus. He stated that local departments have kept local foreign nationals updated on the virus through only and new media platforms like websites, WeChat, Weibo, and e-mail.

The effect of these publications were, however, neutralised when the suspicion with which the Nigerian citizens perceive China took further interesting dimension with media report late April, 2020, that people identified as "a coalition of Nigerian legal practitioners" has instituted a legal action against the Chinese government over the strain the virus has caused Nigerian citizens. The group of twenty-five Nigerians were said to have approached a Nigerian High Court in the Federal Capital Territory seeking $200 billion for losses caused by the pandemic (This Day, 2020). The plaintiff contended that China breached international laws by not informing the World Health Organization about the COVID-19 outbreak on time (Nwachukwu, 2020).

A precursor of the suit was already visible ealier in the pandemic with a former Nigerian Minister for Finance, Oby Ezekwesili (2020) asking China to write-off over $140 billion that "its government, banks and contractors extended to countries in Africa between 2000 and 2017. This would provide partial compensation to African countries for the impact that the coronavirus is already having on their economies and people". Ezekwesili had based her argument for debt cancelation on some details which had indicated China of complicity in the outbreak of the pandemic. According to her,

China should demonstrate world leadership by acknowledging its failure to be transparent on covid-19. Beijing's leadership should then commit to an independent expert panel evaluation of its pandemic response. China and the rest of the Group of 20 countries should engage with the Africa Union and countries to design a reparations mechanism. (2020, para. 14)

In response to such opinion and the reported suit, China adopted a more aggressive communication technique to wage the effect of such action and its viral reportage in the online media space. In its "Statement by Press Secretary of the Embassy of China in Nigeria" published in the Embassy's website, China, in a less diplomatic tone, warned the proponents of the court actions that

Attacking and discrediting other countries will not save the time and lives lost. At this critical moment, we urge that some Nigerian legal practitioners will do more things to enhance mutual trust and help epidemic prevention

and control in both countries, rather than dancing to the tune of a certain country to hype up the situation. (Statement, 2020a, para. 6)

The adoption of what seemed a harsher tune in its reaction in this case may be a diplomatic strategy of wading off any chance of the Nigerian government joining its citizen in the suit. The change of tone was not also surprising as the press statement was directly addressed to the Nigerian citizens, though an indirect warning to their government. China equally used the last part of the statement to hit back at its rival, the US, which it referred to as "a certain country" who the plaintiff dance its tune. This further confirmed that China's coronavirus communications to Nigerians were mainly encoded with the agenda of whittling down the influence of the US-oriented media reportage. The press statement also deployed the denial strategy to out-rightly deny China's culpability in the negative economic growth being witnessed by Nigeria and went ahead to adopt a variant of this strategy by passing blame for the suit and its motivation on "certain" country it accused of incitation.

Effectiveness of China's COVID-19-Oriented Image Laundering Strategies in Nigeria

The effectiveness of the image laundering strategies deployed by the Chinese Embassy in Nigeria could be said to be largely product though not devoid of some seeming counterproductive outcomes. China's changed strategies and retooled its online communication as the pandemic went from the initial stage to the peak of the outbreak. Strategies deployed were largely meant to manage the country's image within specific periods and counter the effect of negative stereotypes and unfavorable conspiracy theories as perceived by Nigeria.

The initial fear that characterized the outbreak of the virus saw many countries shutting down their borders against countries that have large numbers of reported cases of the virus. Not persuaded or convinced by the Chinese argument and criticism against international travel bans at this stage, the Nigeria government went ahead and included the country in the list of country it suspended air travel with. In a motion moved on the floor of the Senate, Chairman of the Senate Committee on Primary Health and Communicable Diseases, Chukwuka Utazi, asked the executive to ban flights from China from landing in the country.

The senator pleaded with the federal government to ignore political and economic considerations and direct the Minister of Aviation, Senator Hadi Sirika, to issue a temporary ban on the Ethiopian Airlines flights to and fro China as the Asian powerhouse remained the main source of COVID-19 (Shibayan, 2020). A further analyses of the senator's speech shows that, apart from his reference to "national interest" such call was motivated by reports on similar decisions taken by countries he referred to as technologically "better and far advanced" than Nigeria.

This seeming determination to protect its people and think independently soon gave way- in a swap of roles-to reactions that evoked the imagery of despondency in 2022 with Nigerian Ambassador to the People's Republic of China, Baba Ahmad Jidda, pleading with China to relax its travel restrictions on Nigerians so as to allow an indigenous airline, Air Peace, maximize the profit of the direct flight approval given to the airline by the Chinese Government.

The Chinese Government had, mid-March 2020, in a seeming excitement that its National Health Commission (NHC) had not reported any new case of the virus, jettisoned its criticism for trans-border movement restrictions and imposed bans on foreigners with valid travel documents in what it said was to prevent second wave of the pandemic through "imported cases" of the virus (Report 2020).

While this situation is not strange to the international business circus, it is smacks of hypocrisy that China managed to turn the table within 12 months going from a country that wanted the rest of the world to allow for free inter border movement to one which has now imposed same restrictions on smaller economy like Nigeria. This goes a long way to show the effectiveness of the online diplomacy deployed by the former at the beginning of the pandemic when it had its back pressed against the wall by negative international reportage.

It is also significant to note that less than one month after the nation-wide condemnation of the arrival of Chinese 15-member medical team in Nigeria, Jidda was reported commending China for assisting Nigeria in tackling the virus. The Minister had stated that China demonstrated evidence of true friendship and solidarity between the two countries. Underling the Nigerian Government acceptance that China, against the popular report by Western media, has been able to contain the virus, the Ambassador pleaded with the country to not only provide further material support to, but also share information on the said successes with the former.

To underline the success of China's image laundering communications, the Director General of the Nigerian Center for Disease Control (NCDC), Chikwe Ihekweazu (2020) agrees with the former on the positive work done to curtail the spread of the virus. Ihekweazu, who was part of WHO delegation invited by China to study the country's fight against the virus reiterated the claim that new cases of the virus had dropped drastically against the numbers being recorded in other parts of the world, noting that the China model is what Nigeria should learn from and try to implement. To reinforce the fact that most of the things reported about China in the media have been bereft of truth, Ihekweazu insisted that it was time for fact, and not fear, using #FactsNotFear hashtag- made popular by the Chinese authorities- to counter fake news and conspiracy theories on the origin of the virus. The significant of Ihekweazu's deployment of this hashtag in his tweet not only serves to transport his line of thought but a significant nation-to-nation solidarity at a time China was battling to have the trust of most of its bi- and multi-lateral connections.

Similarly, the Nigeria government seemed to have also bought into the China's argument that it is helping the rest of the word to fight the pandemic against the image of selfish, profit-conscious stereotype that has already been sold to the government and citizens of the sub-Saharan African country. One of the pieces of evidence in this regard was the speech made by the Nigerian Minister of Health, Osagie Ehanire, while receiving the gift items presented to the country to aid its fight against COVID-19. While addressing the delegation led by Third Secretary Li Guanjie Ehanire, in a symbolic appreciative and exonerative tone, not only commended China for its support for Nigeria at the peak of the pandemic but went to state that country wish to emulate the former, being the only country that understood the understood the language of the virus (2020).

Another dramatic turn of event that could be credited to the China's brand management and damage control was evident in the heated racial issues generated by the discrimination against Nigerians and other Africans in some Chinese cities at the peak of the pandemic. Nigerian Foreign Minister Geoffrey Onyeama had in April, 2020, officially, accused the Chinese authority of discriminating against Nigerians in their country, noting that latter appeared to being subjected to inhumane treatment in public places and stigmatized as carriers of coronavirus. However, shortly after the official China's reactions which was published in the Embassy portal, the Nigerian government backtracked on its initial stance, attributing what it now qualified as "alleged" treatment of Nigerians by Chinese authorities in Guangzhou, the Guangdong province to poor communication between the Chinese authorities and African consulates in Guangzhou, China (Okeke, 2020). The sudden turn in this event, despite the widely published videos only goes to prove the potency of the online crisis management communication deployed by the Chinese government neutralize the negative re-presentation of the country as one with high incidences of racial discrimination.

Chinese digital diplomacy also recorded success is in the area of vaccines acceptance. With the anti-China sentiments and misgivings that dominated the COVID-19 Nigerian media space still fledging, it became very difficult to get the government and the populace buy-in on product manufactured in China. It even became more difficult to market Chinese pharmaceutical product and against the pulse of the Nigerian nation to discuss administration of China-manufactured coronavirus vaccines. This was, however, a gradual shift in this perception which, no doubt is owed significantly to China's image laundering strategies.

The quest to have Nigeria patronize coronavirus vaccines produced by China which has been marketed through subtle persuasive online communications by the Chinese Embassy in Nigeria was also relatively effective. This could be read from the analyses of the Nigeria's Minister of Foreign Affairs, Geoffery Onyeama's press briefing shortly after bilateral meeting with the Chinese Foreign Minister, Wang Yi in January 2021. A significant chunk of the minister's text was used to announce that the Nigerian government had started discussing with their Chinese counterpart on how vaccines produced by the latter could be accessed. Again, the West African reliance on China for fund seemingly played a major role in this decision with Onyeama making reference to the "immense assistance" the country has given to his country, highlight being the donation of protective equipment earlier on in the pandemic.

Having successfully countered the negative narrative over its locally produced vaccines, the Chinese advocacy gained more support from other Nigerians stakeholders and opinion leaders who spoke in support of the product, advising the government to allow China-manufactured COVID-19 vaccine to be administered on Nigerians. The vaccine which had already been approved by for use by over 40 African countries as at February, 2022, had several failed to get the approval of the Nigerian health authorities. Oni (2022), leading this school of thought, advised Nigeria to rethink its non-acceptance of China's coronavirus vaccine which was currently enjoying global acceptance judging by the number of countries that already approved the vaccine.

CONCLUSION

Having explored the relevant literature and thoroughly analyzed the selected online communication samples by the Chinese Embassy in Nigeria, it is pertinent to note that there were clear and subliminal efforts to manage the brand crisis arising from the negative stereotypes that emerged owing to the outbreak of the COVID-19 pandemic. A large chunk of the online posts and publications by the embassy within the period reviewed were encoded as if meant for the global audience but subtly targeted at capturing the attention and sentiment of the Nigerian government and its people.

China's deployment of digital diplomacy in its bilateral engagements in Nigeria was evidently positioned to challenge what is seen as pro-Western sentiment and the resultant anti-China opinions in Nigeria which seem to have been heightened by the closer attachment of the Nigerians to American media spaces and contents. Contents published on the website of the Chinese Embassy in Nigeria were used to counter the influence of these pro-American media and neutralize the extended effects of the said sentiment. To achieve this, China, manipulatively, relied mostly on the positives of its economic transactions with and in Nigeria, with special focus on how the latter had hugely benefited from same.

Putting its positive economic contribution to the Nigeria at the center of its online communications at this period could be seen as a strategy of countering the narrative that it was guilty of counterfeiting, profiteering and exploitation of African countries- a narrative that has only laid credence to the stereotype being capitalized on by some COVID-19 conspiracy theorists to shrink the growing dominance of the country. Ironically, the repetitive emphases on its economic engagements with Nigeria at the peak of the pandemic, runs counter to the image China was trying to put out at the said time with the infusion of business transaction contents in media texts meant to aid transparent health sensitization at a time only staying alive mattered globally was somewhat considered insensitive.

China also took advantage of its media statements it published on its Nigerian embassy website within the period reviewed to flaunt its social aides and medical interventions to the latter. This communication strategy was obviously deployed to portray China as a country who respects human lives and pursues selfless agenda in midst of global health turmoil. Providing further proof of its altruistic agenda, the Chinese Embassy also reiterated its commitments to continue intervening in African health crises, even when - as it repetitively emphasized- other Western country "abandons" the continent.

Evaluating the effectiveness of the China's image laundering strategies at the height of negative conspiracy theories and narratives surrounding the outbreak and the peaking of the coronavirus pandemic comes with a lot of difficulties. This is owing to the fact that implementation of most image laundering strategies are carved to yield long term results. There exist, however, qualitative indices that points to significant successes of the strategies adopted by the Chinese authorities. Judging by reactions of Nigerian government official- consequent of upon the publication of press statement/releases and opinion articles published by its Chinese counterpart, it is obvious that a substantial success was recorded, trying to counter the pro-Western narrative and sentiment promoted by anti-China conspiracy theories.

While the Chinese communication management seems to have achieved reasonable success in influencing favorable reactions and opinion from the Nigerian government, the digital diplomatic efforts do not appear to have had much effect on the way the citizenry perceive China and the role it played in the outbreak of the pandemic. Pro-China sentiments expected to have been wiped up by the favorable disposition of the Nigerian government to the Chinese course have not also rubbed off on the citizens owing to the trust deficit that characterizes government-masses relationship in the country.

REFERENCES

Agency Report (2020, May 26). COVID-19: China places ban on foreigners to prevent second wave of infections. *Independent*. https://independent.ng/covid-19-china-places-ban-on-foreigners-to-prevent-second-wave-of-infections/

Agley, J., & Xiao, Y. (2021). Misinformation about COVID-19: Evidence for differential latent profiles and a strong association with trust in science. *BMC Public Health*, *1*(89), 1–12. https://bmcpublichealth.biomedcentral.com/counter/pdf/10.1186/s12889-020-10103-x.pdf. doi:10.118612889-020-10103-x PMID:33413219

Alanagreh, L., Alzoughool, F., & Atoum, M. (2020). The human coronavirus disease COVID-19: Its origin, characteristics, and insights into potential drugs and its mechanisms. *Pathogens (Basel, Switzerland)*, *9*(5), 331. https://www.ncbi.nlm.nih.gov/pmc/articles/PMC7280997/. doi:10.3390/pathogens9050331 PMID:32365466

Ambassador. (2020*). Ambassador Zhou Pingjian's Exclusive Interview with Punch.* Embassy of the People's Republic of China in the Federal Republic of Nigeria. http://ng.china-embassy.gov.cn/eng/zngx/cne/202005/t20200507_7775332.htm

Benoit, W. L. (2015). *Image restoration theory.* Wiley Online Library/ https://doi.org/10.1002/9781405186407.wbieci009.pub2

Bolsen, T., Palm, R., & Kingsland, J. T. (2020). Framing the Origins of COVID-19. *Science Communication*, *42*(5), 562–585. https://doi.org/10.1177/1075547020953603

Burnard, M., & Richards, A. (2020). COVID-19 and 5G: Biggest cover-up in history? True or false? *INcontext International.* https://www.incontextinternational.org/2020/04/02/covid-19-and-5gbiggest-cover-up-in-history-true-or-false/

Copeland, D. (2012) Public diplomacy, branding, and the image of nations, part iv: Some practical implications. *USC Center on Public Diplomacy.* https://uscpublicdiplomacy.org/blog/public-diplomacy-branding-and-image-nations-part-iv-some-practical-implications

Chow, D. C. K. (2000). Counterfeiting in the People's Republic of China. *Washington University Law Quarterly*, *78*(1). https://openscholarship.wustl.edu/law_lawreview/vol78/iss1/1

De Coninck, D., Frissen, T., Matthijs, K., d'Haenens, L., Lits, G., Champagne-Poirier, O., Carignan, M., David, M., Pignard-Cheynel, N., Salerno, S., & Généreux, M. (2021). Beliefs in conspiracy theories and misinformation about COVID-19: comparative perspectives on the role of anxiety, depression, and exposure to and trust in information sources. *Frontier Psychology*, 1-13. doi:10.3389/fpsyg.2021.646394

Dikotter, F. (1992). *The discourse of race in modern china* (Vol. C). Hurst and Co Publishers Ltd.

Ehanire, O. (2020, March 26). Coronavirus: Nigeria receives donations from China to fight Covid-19. *Premiumtimes.* https://www.premiumtimesng.com/news/more-news/384094-coronavirus-nigeria-receives-donations-from-china-to-fight-covid-19.html

Ezekwesili, O. (2020, April 17). China must pay reparations to Africa for its coronavirus failures. *The Cable.* https://www.thecable.ng/china-must-pay-reparations-to-africa-for-its-coronavirus-failures

Chinese Embassy. (2020). *Fighting Covid-19 China in Action.* Embassy of the People's Republic of China in the Federal Republic of Nigeria. http://ng.china-embassy.gov.cn/eng/zngx/cne/202006/t20200608_7775354.htm

Fleming, D. C. (2014). Counterfeiting in China. *East Asia Law Review, 10,* 14.

FOCAC. (2020). *Joint Statement of the Extraordinary China-Africa Summit On Solidarity Against COVID-19.* Embassy of the People's Republic of China in the Federal Republic of Nigeria. http://ng.china-embassy.gov.cn/eng/zngx/cne/202006/t20200618_7775366.htm

Foreign Affairs Office. (2020). *One world, one fight in solidarity we stand for the building of a community of common health for mankind.* Embassy of the People's Republic of China in the Federal Republic of Nigeria. http://ng.china-embassy.gov.cn/eng/zngx/cne/202004/t20200426_7775307.htm

Foreign Ministry. (2020). *Foreign Ministry spokesperson: We have zero tolerance for discrimination.* Embassy of the People's Republic of China in the Federal Republic of Nigeria. http://ng.china-embassy.gov.cn/eng/zngx/cne/202004/t20200411_7775286.htm

Ihekweazu, C. (2020, March 4). Coronavirus: NCDC DG Chikwe Ihekweazu is sharing what he learned on his trip to China. *Bellanaija.* https://www.bellanaija.com/2020/03/coronavirus-chikwe-ihekweazu-china/

Gauttam, P., Singh, B., & Kaur, J. (2020). COVID-19 and Chinese Global Health Diplomacy: Geopolitical Opportunity for China's Hegemony? *Millennial Asia, 11*(3), 318–340. doi:10.1177/0976399620959771

German Institute. (2018). *New realities in foreign affairs: Diplomacy in the 21st Century.* German Institute for International and Security Affairs. https://www.swp-berlin.org/en/publication/new-realities-in-foreign-affairs-diplomacy-in-the-21st-century

Guangzhou. (2020). *Guangzhou: Facts, solidarity and cooperation.* Embassy of the People's Republic of China in the Federal Republic of Nigeria. http://ng.china-embassy.gov.cn/eng/zngx/cne/202005/t20200504_7775320.htmJidda

Ahmad (2020, May 9). Nigeria appreciates Chinese support in COVID-19 fight. *China Daily*. https://global.chinadaily.com.cn/a/202005/09/WS5eb5fed9a310a8b241154601.html

Hung, K. (2015). *Repairing the "made-in-China" image in the U.S. and U.K.: effects of Government-supported advertising. International Public Relations And Public Diplomacy*. Peter Lang Publishing, Inc.

Jinping, X. (2020). *Keynote Speech by H.E. Xi Jinping President of the People's Republic of China at the Extraordinary China-Africa Summit on Solidarity against COVID-19*. Embassy of the People's Republic of China in the Federal Republic of Nigeria. http://ng.china-embassy.gov.cn/eng/zngx/cne/202006/t20200617_7775363.htm

Kolawole, Y. (2022, February 21). Trade deficit with China worsens with $23 bn imports in 2021.*Vanguard*. https://www.vanguardngr.com/2022/02/trade-deficit-with-china-worsens-with-23bn-imports-in-2021/

Lang, B. (2019). *China and global integrity building: Challenges and prospects for engagement*. Michelsen Institute (CMI).

Lijian, Z. (2020a). *Foreign Ministry Spokesperson Zhao Lijian's Remarks on Guangdong's anti-epidemic measures concerning African citizens in China*. Embassy of the People's Republic of China in the Federal Republic of Nigeria. http://ng.china-embassy.gov.cn/eng/zngx/cne/202004/t20200413_7775302.htm

Lijian, Z. (2020b). *Foreign Ministry spokesperson Zhao Lijian's remarks on Nigerian Foreign Minister Onyeama's address to the press*. Embassy of the People's Republic of China in the Federal Republic of Nigeria. http://ng.china-embassy.gov.cn/eng/zngx/cne/202004/t20200418_7775304.htm

Lijian, Z. (2020c). *Embassy spokesperson's remarks on national security legislation for Hong Kong SAR*. Embassy of the People's Republic of China in the Federal Republic of Nigeria. http://ng.china-embassy.gov.cn/eng/zngx/cne/202005/t20200522_7775339.htm

Liu, L., Huang, Y., & Jin, J. (2022). China's vaccine diplomacy and its implications for global health governance. *Health Care, 10*(7), 1276. doi:10.3390/healthcare10071276

Madrid-Morales, D. (2017). China's digital public diplomacy towards Africa: Actor, messages and audiences, 129-146. Routledge.

Mertha, C. (2005). The politics of piracy: Intellectual property in contemporary China. *Law & Society Review*, *40*(4).

Mitchell, A., & Oliphan, J. (2020 March 18). Americans Immersed in COVID-19 News; Most Think Media Are Doing Fairly Well Covering It. *Pew Research Center* [Blog Post]. https://www.journalism.org/2020/03/18/americans-immersed-in-covid-19-news-mostthink- media-are-doing-fairly-well-covering-it/

Mulualem, M. (2014, April 26). America is coming to Africa: a comparison with China. *AllAfrica*. http://allafrica.com/stories/201404280729.html?viewall=1

Nwachukwu, J. (2020, April 26). Nigerian lawyers drag China to court over COVID-19, demands $200b damages. *Daily Post*. https://dailypost.ng/2020/04/26/nigerian-lawyers-drag-china-to-court-over-covid-19-demands-200b-damages/

Okafor, L. (2020, March 17). Chinese responsible for direct importation of fake products to Nigeria. *Vanguard*. https://www.vanguardngr.com/2020/03/chinese-responsible-for-direct-importation-of-fake-products-to-nigeria-okafor/

Okeke, J. (2020, April 15). Onyeama attributes alleged ill-treatment of Nigerians in China on communication gaps. *Authority*. https://authorityngr.com/2020/04/15/onyeama-attributes-alleged-ill-treatment-of-nigerians-in-china-on-communication-gaps/

Oni, R. (2022). Nigeria Urged to Allow Use of China's COVID-19 Vaccines. *Thisday*. https://www.thisdaylive.com/index.php/2021/11/05/nigeria-urged-to-allow-use-of-chinas-covid-19-vaccines/

Onyeama, G. (2021, January 5). Nigerian Government Begins Talks With China Over COVID-19 Vaccines. *Sahara Reporters*. https://saharareporters.com/2021/01/05/nigerian-government-begins-talks-china-over-covid-19-vaccines

Ouassini, A., Amini, M., & Ouassini, N. (2022). #ChinaMustexplain: Global tweets, covid-19, and anti-black racism in China. *The Review of Black Political Economy*, *49*(1), 61–76. https://doi.org/10.1177/00346446621992687

Page, M. T. (2018). *The intersection of China's commercial interests and Nigeria's conflict landscape*. Special Report by the United States Institute of Peace.

Pearce, K. (2020). Pandemic simulation exercise spotlights massive preparedness gap. *HUB*. https://hub.jhu.edu/2019/11/06/event-201-healthsecurity/

Perthes, V. (2021). Dimensions of rivalry: China, the United States, and Europe. *China International Strategy Review*, *3*, 56–65. doi:10.1007/s42533-021-00065-z

Pingjian, Z. (2020a). *NCP: Time for facts, science and solidarity*. Embassy of the People's Republic of China in the Federal Republic of Nigeria. http://ng.china-embassy.gov.cn/eng/zngx/cne/202002/t20200211_7775252.htm

Pingjian, Z. (2020b). *Ambassador Zhou Pingjian: China's response on COVID-19 responsible, decisive, effective*. Embassy of the People's Republic of China in the Federal Republic of Nigeria. http://ng.china-embassy.gov.cn/eng/zngx/cne/202002/t20200217_7775259.htm

Pingjian, Z. (2020c). *Ambassador Zhou Pingjian: Working together towards a health silk road*. Embassy of the People's Republic of China in the Federal Republic of Nigeria. http://ng.china-embassy.gov.cn/eng/zngx/cne/202003/t20200313_7775276.htm

Pingjian, Z. (2020d). COVID-19: *How China did it*. Embassy of the People's Republic of China in the Federal Republic of Nigeria. http://ng.china-embassy.gov.cn/eng/zngx/cne/202004/t20200403_7775281.htm

Pingjian, Z. (2020e). *Opening Remarks of Ambassador Zhou Pingjian of China at the Embassy's media briefing on fighting 2019 n-CoV*. Embassy of the People's Republic of China in the Federal Republic of Nigeria. http://ng.china-embassy.gov.cn/eng/zngx/cne/202002/t20200204_7775249.htm

Pingjian, Z. (2020f). *COVID-19: Hardship reveals true friendship*. Embassy of the People's Republic of China in the Federal Republic of Nigeria. http://ng.china-embassy.gov.cn/eng/zngx/cne/202002/t20200229_7775268.htm

Pingjian, Z. (2020g). *COVID-19: Economic consequence on China's economy transitory, manageable*. Embassy of the People's Republic of China in the Federal Republic of Nigeria. http://ng.china-embassy.gov.cn/eng/zngx/cne/202003/t20200306_7775272.htm

Pingjian, Z. (2020h). *Remarks of Ambassador Zhou Pingjian at the dialogue held by Centre for China Studies in Nigeria on the theme of "Nigeria-China Cooperation in the context of health emergency: Imperative of joint efforts and collaboration"*. Embassy of the People's Republic of China in the Federal Republic of Nigeria. http://ng.china-embassy.gov.cn/eng/zngx/cne/202002/t20200213_7775255.htm

Pingjian, Z. (2020i). *The Knowledge Center for China's Experiences in response to COVID-19*. Embassy of the People's Republic of China in the Federal Republic of Nigeria. http://ng.china-embassy.gov.cn/eng/zngx/cne/202003/t20200319_7775278.htm

Pingjian, Z. (2020j). *Statement by Press Secretary of the Embassy of China in Nigeria*. Embassy of the People's Republic of China in the Federal Republic of Nigeria. http://ng.china-embassy.gov.cn/eng/zngx/cne/202004/t20200412_7775289.htm

Pingjian, Z. (2020k). *Pandemic: solidarity and cooperation most potent weapon*. Embassy of the People's Republic of China in the Federal Republic of Nigeria. http://ng.china-embassy.gov.cn/eng/zngx/cne/202005/t20200512_7775336.htm

Pingjian, Z. (2020l). *Fighting Covid-19 to build a global community of health for all*. Embassy of the People's Republic of China in the Federal Republic of Nigeria. http://ng.china-embassy.gov.cn/eng/zngx/cne/202006/t20200613_7775360.htm

Pingjian, Z. (2020m). *The strength of China-Africa solidarity in defeating COVID-19*. Embassy of the People's Republic of China in the Federal Republic of Nigeria. http://ng.china-embassy.gov.cn/eng/zngx/cne/202006/t20200625_7775385.htm

Sautman, B. (1994). Anti-black racism in Post-Mao China. *The China Quarterly*, *138*, 413–437. doi:10.1017/S0305741000035827

Shibayan, D. (2020, February 28). Senator asks FG to ban flights to and from China over coronavirus. *The Cable*. https://www.thecable.ng/senator-asks-fg-to-ban-flights-to-and-from-china-over-coronavirus

Shih, C. (2013). Harmonious Racism: China's Civilizational Soft Power in Africa. *Sinicizing International Relations*. Palgrave Macmillan. doi:10.1057/9781137289452_3

Chinese Embassy. (2020a). *Statement by Press Secretary of the Embassy of China in Nigeria.* Embassy of the People's Republic of China in the Federal Republic of Nigeria. http://ng.china-embassy.gov.cn/eng/zngx/cne/202004/t20200430_7775316.htm

Chinese Embassy. (2020b). *Statement by Press Secretary of the Embassy of China in Nigeria.* Embassy of the People's Republic of China in the Federal Republic of Nigeria. http://ng.china-embassy.gov.cn/eng/zngx/cne/202007/t20200708_7775390.htm

Taylor, I. (2007). China's relations with Nigeria. *The Round Table*, *96*(392), 631–645. doi:10.1080/00358530701626073

Tettey, J. (n.d.). Anti-black racism in China, and political economy of asymmetrical power. *Intervention Symposium – "Black Humanity: Bearing Witness to COVID-19" organized by Elaine Coburn and Wesley Crichlow COVID-19.* https://antipodeonline.org/wp-content/uploads/2020/12/7.-Tettey.pdf

Tse, D. K., & Hung, K. (2014). *Chinese firms going global: Their impacts, best practices, and implications.* Cambridge University Press.

Van der Linden, S. (2015). The conspiracy-effect: Exposure to conspiracy theories (about global warming) decreases pro-social behaviour and science acceptance. *Personality and Individual Differences*, *87*, 171–173.

Vincent, D. (2020, April 17). Africans in China: We face coronavirus discrimination. *BBC News*. (2020). https://www.bbc.com/news/world-africa-52309414

Lan, S. (2016). The Shifting Meanings of Race in China: A Case Study of the African Diaspora Communities in Guangzhou. *City & Society*, *28*, 298–318. doi:10.1111/ciso.12094

Vincent, J. (2020, June 3). Something in the air: Conspiracy theorists say 5G causes novel coronavirus, so now they're harassing and attacking UK telecoms engineers. *The Verge.* https://www.theverge.com/2020/6/3/21276912/5g-conspiracy-theories-coronavirus-uk-telecoms-engineers-attacks-abuse

World Health Organisation. (2020). *Origin of SARS-CoV-2.* WHO. https://apps.who.int/iris/bitstream/handle/10665/332197/WHO-2019-nCoV-FAQ-Virus_origin-2020.1-eng.pdf

World Health Organisation. (2020 February 15). Munich Security Conference. WHO. https://www.who.int/dg/speeches/detail/munich-security-conference

Zhu, A. (2020). A lost 'little Africa': How China, too, blames foreigners for the virus. *The New York Review*. https://www.nybooks.com/online/2020/05/05/a-lost-little-africa-how-china-too-blames-foreigners-for-the-virus/

KEY TERMS AND DEFINITIONS

Communication: Communication, in the context of this chapter, is a process through which information is exchanged from an encoder to a decoder.

Coronavirus-oriented: This is a perception, view or phenomenon tailored in relations to or under the influence of the infectious disease caused by the SARS-CoV-2 virus.

Diplomacy: Diplomacy in this involves deployment of communications instruments geared towards country-to-country relationship management.

Image Laundering: This is the professional management of image in ways that can bolster the positive perception of a given country.

Narrative: This is a specific view advanced by a given individual or group meant to form and influence public opinion.

Pandemic: This is an outbreak of diseases that spreads globally, defying known health or medical solutions at the initial stage.

Stereotype: This is a dominant generalised belief about a category of individual or society which are mostly not based on verifiable facts but mainly driven by myths.

Chapter 6
Tweeting to Vindicate the "Marginalised Nigerian Diaspora" in China:
A Content Analysis of Nigeria MFAs' Online Communications

Floribert Patrick C. Endong
iD https://orcid.org/0000-0003-1893-3653
University of Dschang, Cameroon

Eugenie Grace Essoh Ndobo
University of Calabar, Nigeria

ABSTRACT

In a bid to mitigate the COVID-19 pandemic in its territory, the Chinese government embarked on a number of muscular policies right from the early stages of the pandemic. One of such policies was aimed at forcing Africans living in the Guangdong province to accept COVID-19 prevention and containment measures. Subsequently viewed as xenophobic, this policy rapidly degenerated into a diplomatic incident, opposing Chinese and African governments. Through its officials and its diplomats, the Nigerian government in particular condemned the Chinese policies using Twitter among other digital platforms. This chapter seeks to show how Nigerian diplomats deployed Twitter to critique China's perceived xenophobic treatment of Nigerian diasporas in its territory during the early stage of the COVID-19 pandemic. The chapter is based on a quantitative and qualitative content analysis of 12 randomly selected tweets generated by senior staffers at Nigeria's Ministry of Foreign Affairs to denounce the maltreatment of the Nigerian Diaspora in China.

DOI: 10.4018/978-1-7998-8394-4.ch006

INTRODUCTION

The outbreak of the COVID-19 epidemics in China and its subsequent spread in the whole world has caused not only economic damages and heavy losses in human lives, but also far-reaching diplomatic issues involving China and specific nations. Many of such diplomatic issues directly emanated from the perceived "inhuman" ways in which China treated expatriates on her soil, reportedly/apparently in a bid to mitigate and quash the COVID-19 epidemics. One of the communities of expatriates that have suffered such "inhuman" treatment has been African Diasporas – particularly Nigerians living in China. There have, in effect, been various allegations of African expatriates evicted from their homes, stripped of their visas, treated as sub-humans, denied access to hotels /shelters and restaurant and more disturbing, abandoned to suffer and die on the streets of the Chinese city of Guangzhou, all these in the midst of travails provoked by the COVID-19 (Onoja 2020; Sudworth, 2021; Verma, 2020). The above-mentioned allegations have partly been fuelled or inspired by online propaganda, particularly the online contents generated by African bloggers, political activists and citizen journalists.

This state of affairs has, of course, provoked waves of anti-Chinese feelings in African countries (Chu 2020a,b; Chang & Fung 2021; Christensen 2020) as well as in other parts of the world (Gupta, 2020; Adam, 2020). In Nigeria in particular, the government used both official government fora and the unofficial pronouncements of its diplomats to condemn the "maltreatment" of its citizens in China and seek immediate redress. The Nigerian government's reaction to the perceived maltreatment of the Nigerian Diaspora in China can be assessed through a content analysis of Nigeria's MFA's online communications around the incident in April 2020. In this paper, attention is given to two things. First, the paper provides a qualitative content analysis of 12 tweets generated by senior staffers at Nigeria's Ministry of Foreign Affairs to denounce the maltreatment of the Nigerian Diaspora in China and "chronicle" or reveal Nigeria's management of the incident. Second, the paper presents the result of a quantitative analysis of publics' reaction to the 12 tweets. The quantitative and qualitative content analysis has as objective to ascertain both national and international audiences' reaction to Nigeria MFA's response to the maltreatment of the Nigeria Diaspora in China.

FROM "NIGERIANOPHOBIA" TO A DIPLOMATIC INCIDENT: A BACKGROUND TO THE STUDY

In early April 2020, the Chinese authorities announced a campaign aimed at compelling all foreigners living in the Guangdong province to accept "COVID-19 prevention and containment" measures. The campaign included "testing, sampling and quarantine" and was justified by the need for China to mitigate the increase in imported cases of COVID-19 in parts of the country with large communities of foreigners. Although inclusive on paper, the campaign soon became a measure exclusively aimed at African communities which, according to statistics, constitute about 14,000 heads, excluding undocumented Africans. In effect, in guise of applying the policy, the Guangdong authorities began targeting mostly African communities, subjecting their members to forceful testing and quarantine. In the process, the Chinese authorities visited the homes of African residents, testing the latter on the spot or ordering them to take a test at specific local hospitals. A number of Africans were also subjected to self-isolation in houses equipped with surveillance cameras and alarms aimed at monitoring them. In addition to this, the Chinese authorities embarked on impromptus and unannounced inspections at gathering places (notably restaurants and shops) frequented by Africans. Such inspection visits aimed at screening these places and quarantining related personnel and patronisers irrespective of nationality.

A number of sources have predicated the above draconian measures adopted by the Guangdong authorities on rising incidents of imported cased of COVID-19 among African communities (BBC, 2020; 2021; Argentino & Amarasingan, 2020; Seum, 2020); but Human Right Watch (2020) argued that such a thesis had no scientific bases. According to the global human rights observer, most imported cases of COVID-19 were instead Chinese citizens returning from overseas countries. Many of the Africans tested negative for Coronavirus, had absolutely no recent travel history or had not been in contact with unknown people infected with the virus. This line of argument squares well with the contention of Nigeria's Consular in Guangzhou, Razaq Lawal. According to this Nigerian diplomat, maltreated Nigerians "are the people that are living in China, they are not businessmen. That is what I'm telling you when you say money or life. They are here," (cited in Nasiru, 2020).

The Chinese authorities' draconian policy soon provoked waves of xenophobic sentiments against Africans in the Guangdong province. The local Chinese population began to develop a phobia partly fueled by the belief that all African nationalities dueling in the city of Guangzhou are infected, contagious, and very dangerous. This wave of xenophobia complicated the plight of Africans in the town as landlords began to evict Africans from their homes, subjecting the latter to be homeless and more vulnerable. Similarly, local hotels, restaurants, shops and other public establishments banned Africans from using their services. Among the African communities that were the most marginalised featured the Nigerian Diaspora in China, believed by local authorities to be the foreigners with the biggest number of imported cases of COVID-19. In effect, an early investigation had revealed that some Chinese air-transport agencies notably Etihad Air had by their Africa-Guangzhou flights facilitated the importation of African cases infected with the Coronavirus; many of whom were Nigerian citizens. Etihad Air's flight ET 606 from Addis Ababa to Guangzhou alone was, for instance, suspected by March 3, 2020 to have enabled the importation of over 18 cases of infected foreigners in Guangzhou (Niu, 2017). Twelve of these cases were of African origins. Out of these 12 new cases, five were Nigerians and the rest nationals from Kenya, Ghana and Uganda. Worse, some of the cases of Nigerian carriers of the COVID-19 sickness were reported to have eaten at a Nigerian restaurant called Emma Food and whose owner later tested positive for COVID-19. This pushed the Chinese authorities to consider the restaurant as the centre of local infections, and to subsequently screen and quarantine any personnel related to the restaurant and put a special emphasis on Nigerian nationals in the application of their anti-COVID-19 campaign in Guangdong.

China's "muscled" COVID-19 containment measures aimed at African communities soon fueled various social media assisted propaganda campaigns, with some initiated by the victimised Africans as well as staffers at African diplomatic missions in China (Aegentino & Amarasingan, 2020; Zeng, Mike & Schafer, 2021). Indeed, African Diasporas coupled with staffers at some African states' diplomatic missions in China began flooding the cyberspace with troubling versions, images and videos of the ill-treatment of African communities by Chinese authorities and local Chinese population. In their efforts, they directly or indirectly called on their respective countries of origin to take retaliatory or corrective actions against China.

In one of such videos that emerged on April 10, 2020, the Nigerian consular in Guangzhou Rasaq Lawal, is seen publicly censuring Chinese authorities for mistreating Nigerians (Nasiru 2020). In his criticism Lawal stressed the fact that Nigerians were forcefully kept in quarantine beyond the normal 14 days fixed for Chinese citizens. The diplomat also decries the fact that many of his compatriots had their passports seized by the Chinese authorities; all these, not only in violation of international practices and conventions but also in total contradiction to the way Nigeria treats Chinese nationals living in its territory. "This passport belongs to the federal government of Nigeria. In line with international practice, no country has the right to seize the international passport of another country. [...] Why are they seizing Nigerian passports? It does not belong to China. If you seize Nigeria passport, it's like you're seizing Nigeria as a whole. It is not acceptable" Lawal says in the video (cited in Nasiru, 2020).

Other videos that emerged on social media showed Nigerians, among other African nationalities, on the rainy streets of Guangzhou with their possessions. The videos showed them homeless, herded by police officers, stranded, and being discriminated against at hotels, restaurants and being stigmatised by the local population. The ubiquity and exceptionally "distressing" character of these images triggered series of condemnations from Nigerian authorities as well as from other African governments. A number of such condemnations emerged even from the US (Gu, 2021; Jia & Lu, 2021; Krishnan, 2020; Muller, Brazys & Dukalskis, 2021). The chairperson of the African Union Commission Moussa Farki Mahamat in a tweet, claimed to have summoned the Chinese ambassador to the AU to personally discuss the allegations of ill-treatment of African communities in China. Meanwhile, a group of African diplomats in China wrote a letter of complaint to the Chinese Minister of foreign Affairs, censoring the forceful and humiliating quarantines of Africans, the evictions of African nationals from their homes and their stigmatization in China among other troubling issues. The group declared that all these indices of inhuman treatment "in our view, [had] no scientific or logical basis and amounts to racism towards Africans in China" (cited in Campbell, 2020).

Similar to Mahamat and the above group of African diplomats, the governments of Ghana, Uganda and Kenya, launched series of formal and official protests aimed at pressurising China over the maltreatment of their citizens (Campbell, 2020). The maltreatment thus sparked waves of anti-Chinese sentiments in a plurality of African countries with newspapers in Kenya and Nigeria pleading for the prompt government's rescue of Africa Diasporas in China and the immediate repatriation of Chinese living in African countries. Kenya's *Nation* newspaper for instance published an article with the heading "Kenyans in China: Rescue us from hell" while a Kenyan MP called for the repatriation of Chinese living in Kenya.

In Nigeria most specifically, the news of the maltreatment of Nigerian nationals was greeted in political quarters with much resentment. The news sparked serious anti-Chinese reactions from Nigerian politicians and the entire intelligentsia. In an April 14, meeting with the Chinese ambassador to

Nigeria Zhou Pingjian, Nigeria's Minister of Foreign Affairs Mr. Geoffrey Onyema expressed the Nigerian government's "concern" at "allegations of maltreatment of Nigerians in Guangzhou". The Nigerian diplomat stressed the fact that such ill-treatment was "unacceptable" and only warranted immediate Chinese government's intervention. The Chinese ambassador said that Beijing was taking the issues raised by the Nigerian diplomat "very serious" and that China will continuously foster cordial relations with Nigeria (Zhou 2020a,b,c).

In another meeting with the Speaker of Nigeria's House of Representatives, Femi Gbajabiamila, the Chinese ambassador was asked to provide explanations for the ill-treatment of Nigerians as soon as possible. This request pre-empted series of investigations by the Chinese and Nigerian authorities, the results of which eventually led to the resolution of the diplomatic issue.

It should however be noted that the most radical reaction to the ill-treatment of Nigerian came, visibly from the Nigerian House of Representatives. In the heat of the diplomatic row, the House passed a series of motions which, in essence, called for a nationwide audit of the immigration status of all Chinese citizens living in Nigeria and the screening of Chinese businesses in the country for illegal or undocumented persons. The motions aimed at the ultimate repatriation of illegal Chinese immigrants as well as those operating illegal businesses in Nigeria. In addition to the above, the motions called for the evacuation of all Nigerians who were living in China and who wished to return home, including those who went to China for business or who were holders of any form of travel documents, passports, valid or expired visas as well as Nigerians who have tested negative. The motions called for the prompt evacuation of these Nigerians and their being quarantined upon arrival in Nigeria. However, in a 15 April tweet, the Speaker of Nigeria's House of Representative claimed the maltreatment of Nigeria in China and the diplomatic issues between China and Nigeria were all resolved. He wrote:

I'm glad the matter of maltreatment of Nigerians in China has been sorted out between both countries. The Ambassador has communicated his findings and we hope that moving forward communication will be swift and clear and due process will be observed even where there are allegations of wrongdoing by citizens of other countries. It is important that we follow up on this and Nigerians can legitimately go about their business in the People's Republic of China. (cited in Orizu, 2020)

The diplomatic issue between Nigeria and China witnessed a denouement thanks to thorough investigations conducted by the Chinese and Nigerian authorities. These investigations revealed that the video circulated online were of Nigerians who violated the strict measures adopted by China to mitigate the COVID-19 epidemics (*Daily Trust* 2020; Zhou, 2020a,b,c.d; Seum, 2020). The investigations concluded that it is the corrective actions taken by China to deal with cases of violations of the measures which turned out to be misinterpreted by the international community including Nigeria (Zhou 2020c; Seum 2020). In spite of the above results and conclusions of the investigations, the Nigerian government still deplored the fact that, in its attempt to deal with the flouting of the strict measures, the Chinese government did not inform the embassies of the Nigerian flouters. The Nigerian government also censured Chinese government's seizures of Nigerians' passports, qualifying the act as violation of international practices and conventions.

A "TWEETED" REACTION FROM NIGERIAN DIPLOMATS

Africans and Nigerians' reactions to the maltreatment of their compatriots in China were manifested on online platforms through social media discussions, blogging, YouTube messages, and hashtags. This is in part evidenced by the fact that many Africans and Nigerians tweeted under the hashtag #ChinaMustExplain to vent their anger, concerns and fears. Besides this, African politicians notably the Chairman of the African Union Commission, Moussa Faki Mahamat, took to Twitter to either condemn in strong terms what they perceived to be racism against African Diasporas in China; or to reveal the steps being taken by their offices or respective countries of origin to pressure China and ensure a redressing of the situation. In his reaction, Mahamat tweeted on 11 April 2020 that, his office had invited the Chinese Ambassador to the African Union Liu Yuxi, to "express our extreme concern at allegations of maltreatments of Africans in Guangzhou and call for immediate remedial measures in line with our excellent relations," (cited in *Daily Trust*, 2020).

In the same style of communication, the Nigerian Minister of Foreign Affairs tweeted a message merely informing the general public and the Nigerian public in particular of the steps the Nigerian government was taking subsequent upon the alleged maltreatment of Africans in Guangzhou to arrest and discourage in future days, the mistreatment of the Nigerian Diaspora in China. His tweet literally read as Mahamat's own. In effect, he indicated that he "invited the Chinese Ambassador to Nigeria, Mr. Zhou Pingjian to communicate Nigerian Government's extreme concern at allegation of maltreatment of Nigerians in Guangzhou, in China and called for immediate Chinese government's intervention" (cited in Orizu, 2020).

If the tweets of most Nigerian diplomats tended to inform the general public about the various actions taken by the Nigerian government to pressure China, those developed by the Speaker of Nigeria's House of Representative Femi Gbajabiamila, went a little bit farther by posting a video of himself using relatively strong words and pressing the Chinese ambassador on the maltreatment of Nigerians. In his protest, Gbajabiamila actually declared that "it is almost undiplomatic the way I'm talking but it's because I'm upset about what's going". In the video, Ambassador Zhou Pingjian. replied "We take it very seriously."

Thus, some of the troubling images and footage that sparked protests and anti-China online-based narratives across the African continent including Nigeria were the products of Nigerian diplomats' leverage of social media such as Facebook, Youtube or Twitter to censure and pressure the Chinese government over the ill-treatment. This issue will be explained in greater details in the subsequent parts of this essay.

SOCIAL MEDIA DIPLOMACY AND THE LISTENING THEORY

Much has been said or written about the potential of the social media to facilitate or revolutionise the conduct of public diplomacy. In effect, the social media have an incredibly reach and the potential to enable both genuine international information campaigns and harmful political propagandas. With a single tweet for instance, an influential political or religious figure can tremendously influence, if not change the global landscape. Such a political and religious figure can even assess his popularity or that of his/her policies in a country or the world. Conscious of this immense enabling power of digital technologies, most world leaders and diplomats have been using the social media, particularly Twitter as a premier "pipeline of communication" with both domestic and foreign publics. Through social media platforms such as Facebook and Twitter, governments, embassies and policy makers have been able to socialise with both foreign and domestic audiences, thereby practicing what some observers call "good diplomacy" – a type of diplomacy which is about whether countries trust each other and they have good faith in other people's cultures. Presidents such as Donald Trump of the USA have even made social media such as Twitter their strongest political tool.

However, social media diplomacy is bound to follow a number of principles for it to be effective or (more) productive and impactful. One of such principles is listening. The longest-serving Chief Justice of the US Supreme Court and father of American Constitutional Law, John Marshall wrote in the early 1800s that "to listen well is as powerful a means of communication and influence as to talk well" (cited in Sandre, 2012). When making this statement about 200 years ago, Marshall had no idea of the kind of technologies that would be applied in the practice of diplomacy in today's modern times. However, he understood very well the role played by listening in "perfecting" diplomatic communication. In other words, he understood that a good diplomat is one that hears from audiences or communicates bearing in mind that audiences' reactions to his or her communications will help him refine his subsequent diplomatic messages. In his article titled "Conceptualising Public Diplomacy Listening on Social Media", Luigi Di Martino notes that "active listening" coupled with the concept of dialogic engagement is a concrete yardstick by which one can assess successful diplomacy. He explains that "Listening could be narrowly interpreted as a way to implement and readjust a national strategy, or more broadly and ambitiously as an activity that aims to advance international understanding" (Luigi 2020, p.131).

Practicing listening in the conduct of social media diplomacy will therefore mean listening, communicating, being influential and clear in one's message whenever one deploys a social media assisted campaign. This of course, follows from the truism that digital diplomats are bound to communicate their messages to audiences who expect a lively interaction between them and the diplomats. According to Riordan (2012) a digital diplomat who does not apply the concept of dialogic engagement in their practice of social media diplomacy is likely to make his communications boring or irritating to his audiences. As he puts it social media cannot be used only "to flog messages". Most users of social media "expect engagement, they expect other users to listen to them, and respond. Getting your message out on social media, but then not responding to the feedback, irritates other users and is ultimately counterproductive" (p.134).

In the context of social media diplomacy, engagement could, in simple terms, be defined as a situation where there are ongoing interactions between two or more users, involved in an open online conversation. At first sight the above definition may suggest that engagement is easily researchable. However, it must be underscored that the concept is more complex and elusive than the definition suggests. It is in effect difficult to measure engagement in a clear cut manner and relate it to the success of a social media campaign aimed at diplomatic purposes. In spite, of this difficulty, Sandre (2012) highlights two paths a digital diplomat may use to engage their publics abroad and at home. These paths provide a number of indicators one can consider when examining levels of dialogic engagement in diplomats' use of social media for diplomatic goals. Some of the elements mentioned by Sandre in his concept of dialogic engagement include: (a) the digital diplomat should listen to the world around him or her; (b) endeavour to interact with all relevant stakeholders and the general public; (c) inspire others in such a way that is appropriate to his or her message and (d) be personable by producing messages that reflect his or her personality and are not boring among others.

The listening and dialogic engagement theories are old concepts in public diplomacy (as seen in Marshall's 1800 statement). However, recent research has revealed they are yet to inspire or be reflected in many diplomats' use of the social media such as Facebook, Twitter and Instagram in their conduct of public diplomacy. Indeed, many diplomats, embassies and MFAs still tend to mainly enjoy an online presence for its own sake rather than viewing such presence as another tool in the service of broader diplomatic strategies. Others similarly tend to limit their use of the social media to communicating diplomatic messages and consular warning through digital platforms. They have Facebook or Twitter accounts; they tweet, blog and post on WhatsApp; but fail to explore the full potential of the communication tools in their hands. Meanwhile, digital diplomacy should involve listening and dialogic engagement with foreign and domestic publics. This culture will be illustrated in the light of Nigerian diplomats and top politicians' use of Twitter in April 2020 to criticize China's alleged mistreatment of the Nigerian Diaspora.

RESEARCH METHODS

This study is based on the results of two content analyses. The first involved the qualitative analysis of 12 randomly selected Twitter messages generated by senior staffers at Nigeria's MFAs, diplomatic missions in China and Nigeria's House of Representatives, in order to react to allegations of maltreatment of members of the Nigerian Diaspora in China. The qualitative content analysis sought to examine the extent to which the authors of the Twitter messages under study applied listening and dialogic engagement in their use of social media diplomacy. Given the reduced number of tweets deployed by Nigerian diplomats, this contents analysis was done manually.

The second content analysis is quantitative in nature. It involves 798 reactions to the 12 messages generated by Nigerian diplomats. The analysis considered individual posts as the unit of analysis. It also considered a data sheet for the collection of data. This data sheet was structured into four variable labels namely (1) Dialogic engagement between commenters (2) tone, (3) Angle of Criticism, (4) Purpose of posting and (5) Focus. The variable labels are explained as follows:

i. **Dialogic Engagement**: This variable label considered two contents categories namely "yes" for the tweets of commenters who sought to dialogically engage with Nigerian diplomats or fellow commenters and "no" for the tweets of users who sought to do the contrary of the above.

ii. **Tone**: this included three categories namely (a) Anti-China for posts that expressed anti-Chinese sentiments, (b) Pro-China for posts expressing sentiment favourable to China and (c) Neutral for posts generated by commenters who took no side.

iii. **Angle of Criticism**: this variable label focused essentially on anti-Chinese comments. It was structured into three contents categories namely (a) Vindictive for anti-Chinese posts which sought to vindicate the Nigerian/African Diasporas in China, suggesting that harsh or retaliatory actions be taken against Chinese living in Nigeria; (b) Accusatory for posts that essentially sought to tax China with xenophobia and (c) Others.

iv. **Purpose of Posting**: This variable label was divided into three contents categories namely (a) Dialogic for posts in which commenters begged for interactions with diplomats, (b) Non-Dialogic for posts by commenters who did not seek interactions with diplomats and (c) Others.

v. **Focus**: this variable label was divided into three categories of the anti-Chinese comments: (a) Extrapolation, (b) Non-Extrapolations and (c) Noise. Extrapolations were posts that extrapolated from the issue of the maltreatment of Nigerian in China to delve into similar cases of xenophobia by China. The contents category "non extrapolations" was used in reference to comments that stuck to the specific issue of the maltreatment of Nigerians in China. "Noise" represented posts that explored issues that had nothing to do with China, Nigeria or the maltreatment of Nigerians in China. The contents category was used in reference to discussion/comments that tended to seriously depart from the issue under study.

The data collected with the aid of a data sheet was analysed with the use of the Statistical Package for Social Sciences (SPSS) and presented in tables 1-4 below.

PRESENTATION OF FINDINGS

Findings reveal that the great majority (96.61%) of the people who reacted to the 12 tweets under study, did so to dialogically engage with Nigerian diplomats and seek more updates on Nigeria's handling of the situation. Only 3.9% of the commenters did accord importance to the dialogic engagement, as shown in Table 1 below.

Table 1. Frequency distribution for Dialogic Engagement

CATEGORY	YES		NO		TOTAL	
	No	%	No	%	No	%
FREQUENCY	771	96.61	27	3.9	798	100

The findings also indicate that most (95.61%) of the commenters naturally adopted an anti-Chinese tone as shown in Table 2 below. None (00%) of the commenters had positive attitude towards China. Meanwhile only a minority (3.9%) of the commenters were neutral in their reactions to Nigerian diplomat's tweets.

Table 2. Frequency Distribution for the Tone

CATEGORY	Anti-China		Pro-China		Neutral		Total	
	No	%	No	%	No	%	No	%
FREQUENCY	763	95.61	00	00	35	4.39	798	100

Another major finding of the content analysis is that majority (65.39%) of the anti-Chinese commenters simply deployed an accusatory language in their reactions as shown in Table 3 below. Over 26.86% of the commenters. adopted a vindictive attitude towards China, suggesting that the Nigerian government adopts similarly harsh policies against Chinese in Nigeria. A minority of the commenters were neutral.

Table 3. Frequency for Angle of Criticism

CATEGORY	Vindictive		Accusatory		Neutral		Total	
	No	%	No	%	No	%	No	%
FREQUENCY	205	26.86	499	65.39	59	7.75	763	100

Also, the content analysis revealed that a great majority (80.07) of the commenters were non-extrapolative in their messages as they remained focused on the issues that led to the diplomatic incident (see Table 4 below). Meanwhile, only a minority (14.02%) extrapolated and viewed the maltreatment of their countrymen as an opportunity to raise concern about similar anti-Nigerian policies adopted by China in its territory or in Nigeria. Over 5.91% of the commenters generated noise in their reaction to the tweets.

Table 4. Frequency for Focus

CATEGORY	Extrapolative		Non-Extrapolative		Noise		Total	
	No	%	No	%	No	%	No	%
FREQUENCY	107	14.02	611	80.07	45	5.91	763	100

The above mentioned findings are indicative of a number of interesting issues pertaining to Nigerian diplomat's disposition to engage with their followers on Twitters and other interactive social media. The findings also suggest that the largely anti-Chinese reactions of commenters could be a window into the Nigerian imaginary about China during the early stage of the COVID-19 pandemic. The above observation and more will be discussed in the section that follows.

DISCUSSION OF FINDINGS

Discussions in this section will be on two principal issues namely (1) Nigerian diplomats' disposition to dialogically engage with their followers on Twitter and (2) Twitter as a site of anti-Chinese discourse during the diplomatic incident.

A Non-Dialogic Engagement

A visible feature in the tweets of Nigerian diplomats was the absence a dialogic engagement with the public. In effect, the diplomats in most cases simply sought to inform the general public of their appreciation of the situation or of the (radical) measures the Nigerian government had taken or was about taking, to deal with the maltreatment of its citizens in China. They made no concrete efforts at satisfying their interlocutors' request for more updates or dialogic engagement. In his April 9, tweet[1] for instance, Nigerian foreign minister Geofrey Onyema used his Twitter account to merely keep his followers informed about the moves made by his office to pressure the Chinese government towards clarifying and fixing the unfavourable conditions of Nigerians living in China, as shown in Plates 1 and 2.

Figure 1. A Screenshot of Onyeama's Tweet

The tweet has hundreds of replies (see Plate 2). Many of these replies carry requests urging the Minister to throw more light on the outcome of his efforts and the viability of his office' future actions aimed to redress the maltreatment of Nigerian in China. The minister tended to overlook all such requests. He subtly declined the followers' invitations to a dialogic engagement and remained bent on just providing the public with limited insights into Nigeria's plan of action.

Figure 2. A Screenshot of Onyeama's Followin

A similar scenario is observed in the tweets of the Speaker of Nigeria's House of Representative, Honorable Femi, Gbajabjamila. In the latter's April 10, 2020 tweet[2] made in relation to the maltreatment of Nigeria, Gbajabjamila simply informed the public about the fact that he had promptly summoned the Chinese Ambassador to Nigeria, in view of demanding clarification and Chinese intervention to stop the ill-treatment of Nigerian living in China (see Plates 3 and 4). Like Minister Onyema, he avoided any dialogic engagement with his followers.

Figure 3. A Screenshot of Gbajabiamila's Tweet

Gbajabmiala's tweet produced in April 15, 2020[3] to announce the peaceful resolution of the diplomatic crisis is similarly followed by a long thread of replies in which some of his followers requested more information about a number of follow-ups. The parliamentarian totally ignores these requests. Thus, much of Nigerian diplomats' and politicians' tweets in relation to the diplomatic crisis did not include dialogic engagement.

Figure 4. A Screenshot of Gbajabiamila's Followings

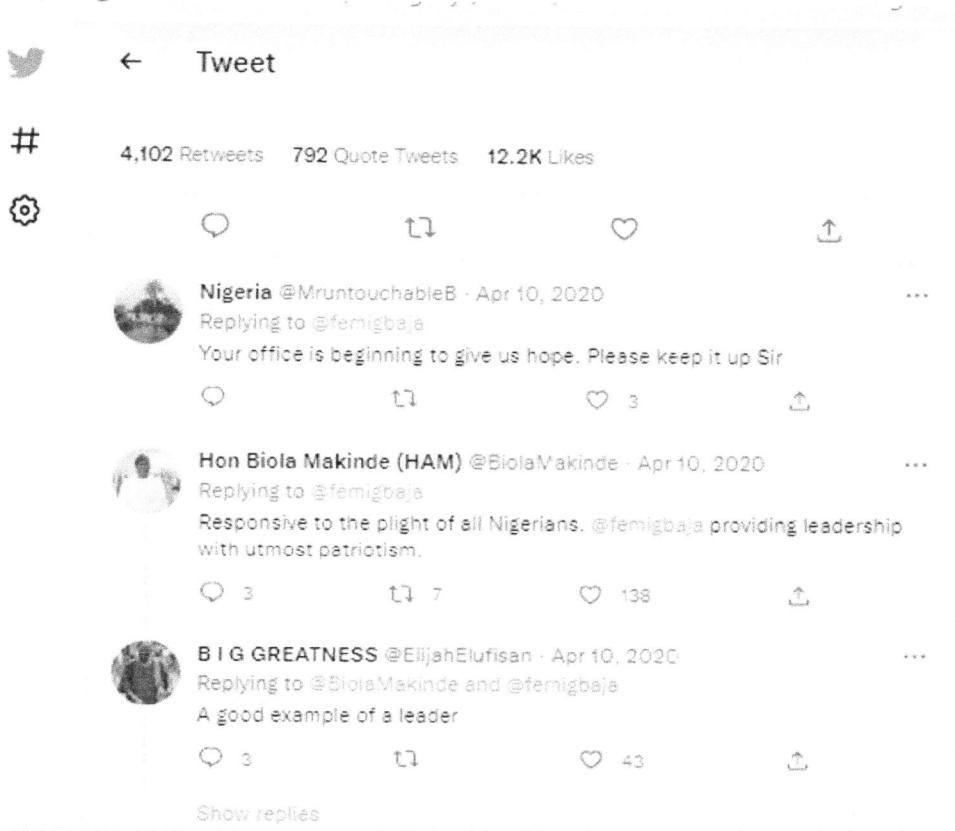

For strategic reasons, a diplomat may be economical with providing information on online platforms, about his government's handling of crisis. However, it remains axiomatic that the inclusion of dialogic engagement in which the diplomat answer the queries of their online publics may go a long way to make their communication and online campaigns more effective and less boring to the online publics. Thus the results of the study reveal that Nigerian diplomats' use of Twitter during the incident was contrary to the prescriptions of Sandre (2012) pertaining to dialogic engagement. Sandre (2012) had actually proposed that diplomats' use of dialogic engagement with the public include the following: (a) the digital diplomat should listen to the world around him or her; (b) endeavour to interact with all relevant stakeholders and the general public; (c) inspire others in such a way that is appropriate to his or her message and (d) be personable by producing messages that reflect his or her personality and are not boring among others.

Twitter as a Channel for Anti-Chinese Protest

The findings reveal that Nigerian diplomats' twitter communications served as a "barometer" of the Nigeria diplomacy's handling of the maltreatment of Nigerian diasporas in China. These communications informed Nigerian about the steps taken by the Nigerian government to investigate and clarify the presumed ill treatment and eventually compel China to adopt corrective or reparatory measures. The Twitter communications however elicited waves of anti-Chinese protest. Nigerian diplomats' followers seized the opportunity to not only appreciate governmental and diplomatic actions taken by the Nigerian government but also to "stage" virtual accusations and protest against the Chinese government.

As shown by the results of the quantitative analysis, the majority of the followers expressed anti-Chinese sentiments while reacting to the tweets of the diplomats. A huge number of these anti-Chinese followers exhibited vindictive attitudes towards China as they called on the Nigerian government to adopt prompt retaliatory actions against Chinese communities living in Nigeria or extrapolated by representing or framing the Chinese maltreatment of their countrymen as a logical continuation of Chinese enterprises' racist policies in Nigeria. It should be noted that the presumed maltreatment of Nigerian citizens in China occurred in a context where a number of precedents pleaded in the disfavour of China. In other words, the diplomatic incident came to aggravate an existing or persisting China's image crisis in Nigeria. In effect, prior to the pandemic and the diplomatic incident, China's image in Nigeria was predominantly negative, though somewhat ambivalent. In the Nigerian popular imaginary, China has for decades been associated with imperialism and neo-colonial exploitation in Nigeria (Endong, 2022; Olander, 2020; Obiezu, 2020; Geerts, Xunwa & Rossouw, 2014). In their numbers, Nigerian newspapers, researchers and opinion moulders had reported cases of Nigeria-based Chinese industries that grossly exploited and maltreated local labourers (Afro Barometre, 2020; Bodomo, 2018; Jackson, 2019; Kobus & Bryan, 2018; ThisDay, 2015; Adewale, 2014). These journalistic narratives and virulent discourses from the Nigerian intelligentsia coupled with many other factors had contributed to tarnishing the image of China even before the Guangzhou incident (Oshodi 2020; Power & Mohan 2010; Quin 2020). Thus, the maltreatment of Nigerian nationals in China simply came to amplify anti-Chinese sentiments in Nigeria. The somewhat vindictive reactions of commenters on Twitter could therefore be viewed as the natural or logical amplification of a sentiment that was born even before the outbreak of the Corona virus.

CONCLUSION

The outbreak of the COVID-19 epidemics in China and its subsequent spread in the whole world have not only caused heavy losses in human lives and economic damages but also far-reaching diplomatic issues involving China and many world countries. Some of these diplomatic issues have directly emanated from the perceived "inhuman" ways in which China has treated expatriates in its territory, reportedly in a bid to mitigate the COVID-19 epidemics. Some of the communities of expatriates that have suffered such "inhuman" treatment have been African Diasporas – particularly Nigerians living in China. There have, in effect, been various allegations of African expatriates evicted from their homes, stripped of their visas, treated as sub-humans, denied access to hotels, banned from shops and abandoned to suffer on the streets of the Chinese city of Guangzhou, all these in the midst of the travails provoked by the COVID-19.

The above mentioned allegations have partly been fuelled or inspired by online propaganda, particularly the social media discussions generated by African bloggers, political activists and citizen journalists. This state of affairs has of course, provoked waves of anti-Chinese feelings in African countries. In Nigeria in particular, the government used both official government forums and the unofficial pronouncements of its diplomats to condemn the "maltreatment" of its citizens in China and seek immediate redress. In this chapter, attention has been given to the manner in which Nigerian diplomats used Twitter to vindicate the marginalised Nigerian diasporas in China.

The chapter argued that Twitter served as a site for anti-Chinese protest involving both Nigerian diplomats and their followers. Nigerian diplomats deployed tweets in which they informed the Nigerian cyber citizens about their government's management of the incident. On Twitter, Nigerian diplomats and government officials denounced the maltreatment of Nigerian communities living in China, thereby eliciting huge anti-Chinese sentiments from their followers. Followers mainly used the diplomatic incident as a departure point to critique China's policies in both Nigeria and in its territory. Followers' reactions to the tweets revamped or resuscitated a plurality of negative stereotypes of China. Some of these stereotypes include imperialism, neo-colonialism and racism. Thus, Nigerian diplomats' use of Twitter during the diplomatic incident inspired Nigerian cybernautes' reassessment of China's COVID-19 related policies as well as age-old stereotypes of China in Nigeria.

It goes without saying that the corpus considered for this study (12 tweets) is relatively reduced. It is also clear that a larger corpus would have enabled a clearer view of both Nigerian diplomats' use of Twitter and followers' attitudes towards Chinese COVID-19 related policies. However, this study provides fresh findings on the Nigerian popular imaginary about China. It throws light on how Nigerians appreciated the diplomatic actions taken by their government to solve the presumed racist Chinese policies aimed at Nigerians.

REFERENCES

Adam, G. (2020). China's failed pandemic response in Africa. *Foreign Policy Essay*. https://www.lawfareblog.com/chinas-failed-pandemic-response-africa

Adewale, M. P. (2014). Nigeria's China connection. *New York Times*. https://www.nytimes.com/2014/05/08/opinion/majapearce-nigerias-china-connection.html

Andreas, S. 2012. Social Media diplomacy: the rules of engagement. *Diplo*. https://www.diplomacy.edu/blog/social-media-diplomacy-rules-engagement

Argentino, M. A., & Amarasingam, A. (2020). The COVID conspiracy files. A Technical report. New York: GNET: Global Network on Extremism and Technology.

Barometre, A. (2020). *Africans' perception about China: A sneak peek from 18 countries*. Afro Barometre.

BBC News. (2020). Coronavirus: Chinese chief for Kano say make pipo no fear dem. *BBC News Pidgin*. https://www.bbc.com/pidgin/tori-51702073

BBC News. (2021). Covid origin: Why is the Wuhan lab-leak theory taken seriously? *BBC News* https://www.bbc.com/news/world-asia-china-57268111

Bodomo, A. (2018). Is China colonising Africa? In S. Raudino & A. Poletti (Eds.), *Global economic governance and human development* (pp. 122–135). Routledge. doi:10.4324/9781315169767-7

Campbell, J. (2020). Despite new China-Africa tension, Beijing has pivotal role to play in Africa's COVID-19 recovery. *Foreign Relations*. https://www.cfr.org/blog/despite-new-china-africa-tension-beijing-has-pivotal-role-play-africas-covid-19-recovery

Chang, C. K., & Fung, A. Y. H. (2021). From soft power to sharp power: China's image in Hong-Kong health crises from 2003-2020. *Global Media and China*, *6*(1), 62–76. doi:10.1177/2059436420980475

Christensen, T. J. (2020). *A modern tragedy? COVID-19 and US-China relations*. Foreign Policy.

Chu, M. (2020a). China will defeat the Coronavirus. Vanguard.

Chu, M. (2020b). Nigeria's support helpful to fight Coronavirus. Vanguard.

Daily Trust. (2020). Editorial: Maltreatment of Nigerians in China. *Daily Trust*.

Di Martino, L. (2020). Conceptualising Public Diplomacy Listening on Social Media. *Place Branding and Public Diplomacy*, *16*(2), 131–142. doi:10.105741254-019-00135-5

Dong, N. (2017). 'Unequal Sino-African Relationships': A Perspective from Africans in Guangzhou. In Y.-C. Kim (Ed.), *China and Africa: A New Paradigm of Global Business* (pp. 237–259). Palgrave Macmillan.

Endong, F. P. C. (2022). Re-branding China's battered image in Nigeria amidst the COVID-19 pandemic. A qualitative analysis of Chinese diplomatic communications. *Journal of BRICS Studies*, *1*(1), 26–40. doi:10.36615/jbs.v1i1.615

Geerts, S., Xunwa, N., & Rossouw, D. (2014). *Africans' perceptions of Chinese business in Africa. A survey (Globethics net Focus No. 18)*. Ethics Institute of South Africa.

Gu, F., Wu, Y., Hu, X., Guo, J., Yang, X., & Zhao, X. (2021). The role of conspiracy theories in the spread of COVID-19 across the United States. *International Journal of Environmental Research and Public Health*, *18*(7), 1–14. doi:10.3390/ijerph18073843 PMID:33917575

Gupta, A. (2020). Clashes over COVID-19 aid in Nigeria. China Africa Project. *China African Project*. https://chinaafricaproject.com/student-xchange/clashes-over-covid-19-aid-in-nigeria/

Human Rights Watch. (2020). *China: COVID-19 Discrimination against Africans: Forced quarantine, evictions and refused services in Guangzhou.* Human Right Watch.

Jackson, S.F. (2019). Two distant giants. China and Nigeria perceive each other. *European Middle Eastern and African Affairs*, 40-74.

Jia, W., & Lu, F. (2021). US media's coverage of China's handling of COVID-19: Playing the role of the fourth branch of government or the fourth estate? *Global Media and China*, 6(1), 62–76. doi:10.1177/2059436421994003

Kobus, J., & Bryan, R. (2018). *China's impact on the African renaissance: the baobab grows.* Palgrave Macmillan.

Krishnan, A. (2020). The COVID-19 pandemic is China's biggest crisis since Tiananmen, says Richard McGregor. *The Hindu.* https://www.thehindu.com/news/national/coronavirus-the-covid-19-pandemic-is-chinas-biggest-crisis-since-tiananmen-says-richard-mcgregor/article31438204.ece

Muller, S., Brazys, S., & Dukalskis, A. (2021). *Discourse wars and Mask Diplomacy: China's global image management in times of crisis (Working Paper 109)*, Dublin: AIDDATA.

Nasiru, J. (2020). Video: Stop seizing the passports of our citizens – Nigerian diplomat tackles Chinese officials. *The Cable.* https://www.thecable.ng/video-stop-seizing-the-passports-of-our-citizens-nigerian-diplomat-tackles-chinese-officials

Obiezu, T. (2020). Coronavirus concerns spur Nigerian authorities to close Chinese market in Abuja. *Voice of America.* https://www.voanews.com/science-health/coronavirus-outbreak/coronavirus-concerns-spur-nigerian-authorities-close-chinese

Obiorah, N. (2008). *Rise and right in China-Africa relations: (SAIS) Working paper in African series,* Washington: School of advanced International.

Olander, E. 2020. Nigeria's unprecedented censure of China. *The China-Africa Project.* https://chinaafricaproject.com/analysis/nigerias-unprecedented-censure-of-china/

Onoja, A. (2020). How China lost Nigeria. *The Diplomat.* https://thediplomat.com/2020/08/how-china-lost-nigeria/

Orizu, U. 2020. COVID-19: Maltreatment of Nigerians in China resolved, says Gbajabiamila. *ThisDay*. https://www.thisdaylive.com/index.php/2020/04/15/covid-19-maltreatment-of-nigerians-in-china-resolved-says-gbajabiamila/

Oshodi, A. G. T. (2020). Nigeria and China: Understanding the imbalanced relationship. *The African Report*. https://www.theafricareport.com/29060/nigeria-and-china-understanding-the-imbalanced-relationship/

Power, M., & Mohan, G. (2010). Towards a critical geopolitics of China's engagement with African development. *Geopolitics*, *15*(3), 462–495. doi:10.1080/14650040903501021

Quin, D. (2020). Nigerian living near a major Belt and Road project *Pew Review*. https://www.pewresearch.org/fact-tank/2020/04/23/nigerians-living-near-a-major-belt-and-road-project-grew-more-positive-toward-china-after-it-was-completed/

Seum, S. (2020). Guangzhou: Facts, solidarity and cooperation. Leadership.

Shaum, R. (2016). Executive Summary: Digital diplomacy 20: Beyond the social media obsession. In S. Riordan (Ed.), *The strategic use of digital and public diplomacy in pursuit of national objectives* (pp. 2–4). Forcir Pensament.

Sudworth, J. (2021). Wuhan marks its anniversary with triumph and denial. *BBC*. https://www.bbc.co.uk/news/world-asia-china-55765875

ThisDay. (2015). China–Nigeria ties as a framework for attaining UN Sustainable Development Goals. *This Day*. http://allafrica.com/stories/201511162091.html

Umejei, E. (2015). China's engagement with Nigeria: Opportunity or opportunist? *Africa East-Asian Affairs*, *3*(4), 54–78. doi:10.7552/0-3-4-165

Verma, R. (2020). China's diplomacy and changing the COVID-19 narrative. *International Journal (Toronto, Ont.)*, *75*(2), 248–258. doi:10.1177/0020702020930054

Zeng, J., Mike, S., & Schafer, S. (2021). Conceptualising "Dark Platforms". Covid-19-related conspiracy theories on 8kun and Gab. *Digital Journalism*, *41*(2), 1–23.

Zhou, P. (2020a). China's response on COVID-19 responsible, decisive, effective. Leadership, January 29, edition, p.44.

Zhou, P. (2020b). COVID-19: Hardship reveals true friendship. People Daily, (February), p.22

Zhou, P. (2020c). COVID-19: How China did it. Leadership, (April), pp. 22

Zhou, P. (2020d). NCP: Time for facts, science and solidarity. New Telegraph, (February), p.30.

ENDNOTES

[1] https://twitter.com/GeoffreyOnyeama/status/1248348204950831104

[2] https://twitter.com/femigbaja/status/1248698266889457664?lang=en

[3] https://twitter.com/femigbaja/status/1250383003748106240?lang=en

Chapter 7

Using Digital Diplomacy in the Context of the COVID–19 Pandemic:
The Indian Experience

Hameed Khan
https://orcid.org/0000-0002-6801-4711
Guru Ramdas Khalsa Institute of Science and Technology, Jabalpur, India

Kamal Kumar Kushwah
Jabalpur Engineering College, India

ABSTRACT

Epidemiology and scale management has been an ongoing project of governments worldwide, including India, focusing on reducing the spread of the virus and reducing the social and financial damage caused by the virus. With the closure of health infrastructure and outbreaks affecting health care workers, there is a need for a cooperative approach to the management of Covid-19 and greater involvement of public enterprises to lead government efforts to control the epidemic. The latest digital technology plays an essential role in monitoring the situation closely, assisting the government, and utilizing high-risk public organizations. Diplomacy—whether regular or non-digital—is affected: seminars and seminars are transmitted to Zoom and various video platforms; governments and international organizations have worked closely with social media, digital technology, and content boards to combat the information that is not in line with COVID-19; foreign policy actors in public and private sectors who have tried to meet new audiences, each online and offline.

DOI: 10.4018/978-1-7998-8394-4.ch007

INTRODUCTION

The world is fighting the COVID-19 epidemic. Along with various nations, India has forced the lockdown to stop the collection of corona diseases. Social isolation is the only solution until a corona vaccine is available. The implementation of strict closures through the authorities in all parts of the country and at the same time performing their regular duties is no longer an easy challenge for us with a large population of 419.80 people per square foot. We can monitor, control the crowd, reduce illegal movement, and locate hotspot circuits using technology. The construction of four specialized units: drone hiring, traffic for surveillance cameras, cell phone tracking, and gas distribution to authorized people using the cell gadget to reduce illegal movement. This chapter emphasizes the benefits of using a technology-driven model as scientific gadgets are advised to be used for monitoring. Authorities may not be as small as the virus as these devices operate or are blocked in an enclosed and out-of-reach area. And the use of real-time statistics using monitoring tools will give us the highest accuracy and allow us to take immediate action in the event of a violation of any regulation. Immediate action leads the province to an area that can control the occurrence of COVID-19 and provide assistance without delay.

Coronavirus infection (COVID-19) does not reduce it without a vaccine, but it does contribute to human immunity, and the economic system can be reduced through advanced disease technology. The introduction of new programs and strategies has been developed to prevent COVID-19 risk effectively. Modern research incorporates the function of intelligent technology in decreasing the unfolding of COVID-19 thru the simultaneous focal point on creating digital technologies. The AI techniques and technologies framework highlights comprehensive data analysis, predicting air pollution threats, presenting scientific assistance, and evaluating diagnostic results. Apart from this, technological advances in masks and sensory technologies throughout the epidemic have been developed that encompass strategies such as 3D printing and visual acuity, respectively. In addition, the strengths, weaknesses, opportunities, and potential threats posed by the complex implementation of this science are also included in detail (Jiang, 2021).

Coping with the COVID-19 pandemic is an assessment for more and better digital diplomacy as the world goes digital and practices social distancing, vaccination, and digital technology transformation. Digital diplomacy needs to focus more on mutual relationships between national and international states based on common interests. The primary focus on India's regional diplomacy and its relevance in the context of the COVID-19 pandemic assesses India's reach in the region with a focus on health, digital and economic diplomacy in light of the ongoing pandemic situation. It also analyzes the challenges India faces, considering the response of its neighbours to its regional policies.

DIGITAL GRAPHS AND TRANSFORMATIONS HAVE HAD AN IMPACT ON THE ENTIRE EPIDEMIC

The disease has built the year 2020 in many ways. Diplomacy - whether regular or non-digital - is affected: conferences and seminars are transferred to Zoom and various video platforms; governments and international organizations have worked closely with social media and content boards to combat the information that is not in line with COVID-19; foreign policy actors in public and private sectors who have tried to meet new audiences, each online and offline (Abbas et al., 2021).

While the epidemic has affected every aspect of our daily lives - now not just political and communications - some issues have been raised over the past year, many of which are phases of this review:

- Technological geopolitics
- New platforms, new tools, and new influencers
- New challenges and reporting problems

The Rapid Transformation to The Internet and Social Media Requires Fundamental Changes

Online structures need to be improved to provide significant stability and protection of negotiations; Face-to-face demands to be transformed into new internet power; many insurance policies that have been in place for centuries need to be reviewed. The new 'hybrid diplomacy' that incorporates regular face-to-face meetings on online topics, appropriately as ad-hoc online meetings, is overgrowing. Online conferences, conferences, and various activities are now in place as COVID-19 spreads (and spreads again), forcing people to abide by anti-social laws. Hundreds of jobs have transitioned from web page to queue for the first time (Ahmad & Murad, 2020). A new concept, the definition of digital diplomacy, is developed and used interchangeably with phrases such as e-diplomacy and tech diplomacy. It is often understood as using the Internet and Information and Communication Technology (ICTs) to advance discussion agendas.

"When it happens in the world, it happens on Twitter," recalls Agung Yudha from Twitter, while showing the summaries of the world's leaders working on Twitter, from the most mobile and influential to the most connected and most connected. It used to be fun to hear how governments and strategists now count the number on a tweet as the arrival of "short and happy" memories to make their positions in specific policies ready for data sharing and updates in real-time. Digital diplomacy has gone beyond simply working on social media. A key element of his presentation should be how the media should act as a voice exchange tool and how digital leadership for thousands of users should contribute to expanding embassy services related to safety and migration, informing strategic discussions, and monitoring development assistance programs (Kadam & Atre, 2020). The digital revolution has disrupted how the global community communicates, communicates, and exists. The diploma, in particular, is the only region already affected and will continue to emerge as strategists set out the equipment and methods that appear. COVID-19 has forced the international community to use digital chat channels. Overcrowded meetings, coffee talks, shoulder taps, and overcrowding have replaced scheduled meetings, emails, online chats with home offices. This transformation of digital structures has opened up new opportunities while highlighting more than a few challenges. Digital diplomacy can subsequently create an integrated and participatory strategy for word exchange and decision-making, or not. Respect for neighborly quality prevents the development of the best working relationships you wish to succeed. At the same time, the high explosion of digital communications is likely to facilitate informal change while maintaining our genuine commitment. Meeting online reduces barriers to participation such as visas and travel expenses. It provides many officials and private visitors opportunities to participate in the discussions to improve the conversation. Even joint meetings present new opportunities, as it takes long-term care to ensure that at least 40% of attendees participate remotely to ensure stability and free participation. It is important to remember, however, that digital segregation is accurate. Access to scientific and web a connection profoundly affects others and people's ability to watch meetings (Madan Gopal, 2020).

There are many types of data accessible to the public, from which Dos Digital Diplomacy Dos and Don'ts are drawn to apply the present and to think of ways to build the future.

- Open the online room at least 5 minutes before the deadline to allow members to enter and ensure that the first time is up to date.
- Regularly announce the time, the privacy rules, and the participation process - this can be done orally or by repeating a minor key in the dialog box.
- Encourage those who are often on the sidelines to talk or ask questions
- Determine gender/age/race range in activities with more than one speaker
- Imagine a conversational discussion (critical points summarized by the counselor, received at the beginning and at the same time)
- Collect and share anonymous information about people of different genders, ages, and locations Give everyone and many a way to find out who is in the actual meeting so that donors can work together
- Ensuring that the selected platform participates in non-discriminatory discussions between members to facilitate discussions and collaboration
- Consider using multiple streams of simultaneous access streams (e.g., Facebook, YouTube, video conference service)
- Provide additional languages when electricity is available (other programs, such as the Zoom institution-subscription institution, including
- Share tips for using the meeting platform ahead of time
- Use slides/shows when audible (it can be challenging to keep your face focused for a long time)
- Use a clear camera
- Post logical facts, statistics, and online links for chat activity
- Replace your involvement when installing the platform
- Keep your video off, without speaking, when conferences are large enough to help people across the border participate easily.
- Keep your microphone quiet, without speaking
- Encourage questions about the discussion activity
- Repeat aloud, questions are asked throughout the conversation
- To your knowledge, make sure that most participants ask questions
- Ignoring access to people with disabilities
- Allow a few circuits or a small group of people to control the conversation
- Reduce public participation
- Long-distance travel without breaks
- Assume that people will be able to see in the element on the slide
- Use small fonts on slides

- Use very long Internet links
- Forget location monitoring to find recordings (if any)
- Take screenshots unless it's public time

Above all, it is essential to consider that the occasional digital dialogue of COVID-19 brings the conference from conference rooms to private workplaces or residences and be aware that delegation care responsibilities, often hidden in the scientific world, will be more visible. This situation requires flexibility, perseverance, and understanding for all stakeholders.

Challenges of Covid-19 On Community Organizations Over

Businesses worldwide have tested the foundations, nonprofits, community organizations, and individuals to find out how the COVID-19 epidemic affects them. The findings of these tests provide facts to help the quarter address this issue, strengthen solidarity, and value representation. Representatives of nonprofit organizations, donors, and community-based organizations (CSOs) participated in the surveys (Rick, 2020).

We are assessing the impact of the epidemic on organizational performance.

- Identifying how organizations respond to these challenges.
- Identifying opportunities and lessons learned.

The COVID-19 epidemic has exacerbated and exposed many of the global challenges facing communities. Still, in addition, it highlighted the weaknesses, challenges, and opportunities within the philanthropic and nonprofit sectors. The survey answers help us to see what those are. Our analysis below falls into five categories:

- Financial challenges
- Challenges away from work
- New ways of dealing with the problem
- The importance of client assistance in building resilience
- A variety of perspectives

Organizations Find It Difficult to Work Remotely

Some groups are fighting increasing pressure and restricted technological power as they strive to operate remotely. After fundraising and economic planning, top leaders of WINGS member organizations worldwide have set "welfare and self-care and team" as their top concern of 0.33. More than 40 percent of Brazilian companies reported that their employees were confused and overworked when looking for their partners. There is a lot of work for many because of COVID-19, and organizations find it challenging to help their employees. More than 70 percent of respondents in the May CAF America survey saw workers' salaries as their preference for labor costs (Volodenkov & Pastarmadzhieva, 2020).

In some cases, staff recruitment, benefits, and salaries are reduced as practical measures. Many businesses have discontinued some of their online services or other remote facilities. Still, some are unable to transfer their offerings online due to lack of required technology, lack of technical capacity by staff and beneficiaries skilled to use online platforms, or honestly because their systems do not fit well with remote formats. CAF has learned about Russia has determined that only 26 percent of Russian-based NGOs have long-term employment applications (Manor, 2021). Nonprofit Institute found that more than half of San Diego-based nonprofits need scientific access to manage remote operations.

Organizations Are Finding New Ways to Help Their Communities With a Different Response To This Problem

Organizations use new ways to support their responses, such as giving or additional donations from neighboring communities, financial savings at canceled events, and building relationships. They also use physical activity programs, play an essential role in sharing facts with nearby neighborhoods, and change their schedules to be patient over time.

About 1/2 of all CSOs tested worldwide have brought new response services directly to the epidemic. Examples include disseminating accurate facts about the novel coronavirus, delivering food and resources to families in need, abundant water supply training, food, housing, energy, waste management, and psychological and psychological support. While the Communications Network is asking for nonprofits and U.S. bases, "Has your employer set up an internal team to navigate the COVID-19 crisis?" 69 percent responded that they did. These operating companies include senior executives and managers, communications staff, and software and staff.

Organizations report exploring and building relationships because of the problem, each response linking and helping other businesses stay active. According to a Mexican Center for Philanthropy study, about 20 percent of CSOs in Mexico are already using collaborative methods. In addition, there is the experience of hope that high-quality trading can be done through forced change of problem. Forty-five percent of respondents in a survey conducted by @AfricanNGOs and EPIC-Africa agreed that teams would come out better and faster after the epidemic. More than half of them wrote about their recent experience incorporating monitoring and evaluation frameworks. In addition, they hope that the crisis will force donors to reconsider their power and move forward with strategies that give impetus to the long-term sustainability and independence of African CSOs. Peace Direct learned about the peacemakers of the neighborhood and saw that they hoped that this time was once an opportunity to strengthen brotherly love for society and to include more practices for conversion and resilience.

A Few Integrated Acquisitions That May Be of Interest

A nonprofit leader is essential in gender issues regarding transparency about future funding from essential donors. Big donors are more likely to talk about male-led nonprofits and less likely to talk to female-led donors about how they will show them in the future. ("Donor Support in the wake of COVID-19 disease," Active Distribution Center)

Younger generations format a more extensive supply than others. Another 46 percent of millennials say they will increase donations in response to the epidemic, in contrast to 14 percent of Baby Boomers and 25 percent of Gen X. Those who say they will reduce their contributions to charities are more concerned about the economy. ("COVID-19 and philanthropy: How donors react to changes between diseases," Fidelity Charitable)

Since the beginning of the disaster, few people have managed and spent money. Previous research has shown that more than 50 percent of donations are made financially in addition to technological advances. About 63 percent of charities say they can take some digital donation through their websites, online platform, or wireless donations. Another 23 percent, however, said they would no longer be able to access digital donations, which is depressing given the low numbers of people in control across the epidemic. Not all governments have supported or identified the efforts of CSOs, nor have they all provided guidelines to reduce the impact of the epidemic on the overall performance of CSOs and application activities. About 72 percent of African CSOs surveyed feel that governments have failed to recognize and use local skills, experience, and networks. It has undermined planning, collaboration, and sustainability across the country. Nonprofits are staffed, and the availability of volunteers is usually due to child care or support. More than 1/2 employees are already facing this problem, and some are relying on that as we enter the new daily and ongoing closures. When asked, "What surprised you the most about your help in recognizing the onset of novel coronavirus?" The top three answers were: "How the business has adapted to the new environment," "Staff capacity," and "Volunteer Volunteerism." Neighbor is key to getting COVID-19 (Turchetti & Lalli, 2020).

- The COVID-19 epidemic has identified systemic errors and significant inequalities.
- Communities play an essential role in reviving the global and financial crisis.
- In the aftermath of a crisis, public companies can act as defense attorneys and rely on authority.

DIGITALLY OPPORTUNITIES FOR CIVIL SOCIETY ORGANIZATIONS IN COVID-19 EPIDEMIC CONTROL

The novel coronavirus that has plagued the world has been on its knees in the wake of a significant frustration among health workers. It has made the global economic system worse than before, sadly far from over (Shankari et al., 2020).

Despite the most substantial closure, India is ranked as the second-highest number globally after the United States. Although we have seen a slight decline in the rate of progress, as the festivities approach the nook and the limits of mitigation, we need to cross with a warning if we are to overcome this difficulty. We have looked at the tested results of ongoing testing, tracking, and segregation throughout India, and it is safe to say that it is one of the best and most natural ways the virus can be detected and contained (Mullard & Aarvik, 2020).

Managing the epidemic and scale has been a progressive project of governments worldwide and India, with a focus on reducing the spread of the virus and reducing the social and financial harm caused by the virus.

With the banning of health infrastructure and outbreaks affecting health care workers, there is a need for a cooperative approach to the management of Covid-19 and higher involvement of public enterprises to guide government efforts to control the epidemic. South Korea has been applauded for quickly managing the epidemic. The country's civil society plays an essential role in monitoring the situation closely, assisting the government, and implementing high-risk public organizations.

A Position of Community Mobilization in The Management Of The Covid-19 Epidemic

In the past, promoting neighborly assistance through CSOs has helped governments manage and eradicate polio, measles, rubella, and smallpox and has helped fill gaps in access to management.

The most recent example that can show us is that civil society organizations (CSOs) are coming up with new ways to fight Covid-19 and reach higher levels of risk in Southeast Asia. Thai CSOs made illegal data and fought two wars, notably 'infodemic' (as referred to using WHO officials) and epidemics. It passed information anonymously to Thai disease control authorities. When enough data is collected, the algorithm explores a new novel coronavirus hotspot that can create and sing humans appropriately.

In Bangladesh, about 200 NGOs have provided financial support to local and clinical NGOs and food aid. They then distributed hygiene items to the needy and spread information about the disease. In the 1970s, India was often celebrated for eliminating smallpox with the help of its civil society organizations, with hundreds of health care workers and one WHO-trained local worker going from house to house in the U. S. It included 100 million homes in 575,721 villages and 2,641 cities. Their work bore fruit when a deadly plague was often eradicated from India in 1977. Similarly, tireless efforts by NGOs and CHOs to stop novel coronavirus will be re-recorded as soon as the epidemic is over.

It also called on CSOs to stand up and work tirelessly with the government, especially in closing and closing after ensuring that communities have the right to access critical services. However, there are opportunities for CSOs to play an increased role in managing Covid-19 inpatient care, encourage the public to behave responsibly, promote hygiene, and connect with the people with whom they communicate.

The Need for Cooperation

As we enter the essential phase of the epidemic, our involvement with CSOs will ultimately hold the key to reducing persecution. Ensuring handwashing, respiratory testing, contracting and tracking contacts, accurate identification and critical segregation of the epidemic management base, and disseminating information through these private organizations helps boost self-esteem and reduce stigma. In addition to technological advances in contractual recognition, use is limited. The operation of that technical interface will require healthy involvement and compliance at the community level, which can be done with the help of these organizations.

Making the first step is using a working staff team and encouraging staff growth and satisfaction can help solve the current problem on a large scale. Earlier this year in May, the state-owned Empowered Group 6 mobilized a network of 92,000 CSO / NGOs to mobilize their resources to assist national governments and regional. Administrators find tropical areas and send volunteers to provide essential services to participants and build corona considerations on government donations.

The Impact of Covid-19 On Social Media Use

In addition to unconventional purchases, desirable shows, and full-scale, in-store promotions, ads on novel coronavirus technology have gone digital.

With store purchases suspended and banned, social media has always been a gateway to consumers' homes to see that March. How can products bring you closer to revenue through this increased access and confidence in e-Commerce?

"All the signs indicate that there was an exchange," Murphy said. "We see positive engagement in Facebook (owners) groups, as people tend to share their experiences and interact with their peers through the same things. As a product, it's high time to establish a social media marketing strategy, like establishing strong links" (Depoux et al., 2020).

Shopping During Covid-19

While the media has performed a primary position in the development of the buying trip, there will still be T.V. advertising - although the current generations of hastily developing markets are changing the face of the world, particularly considering the introduction of COVID -19.

Social media advertisements have more product availability as buyers enhance their show time throughout the lock. They even went thru regular nature searches and T.V. commercials. The Brands recently spoke on social media, noting that 47% of shoppers create surroundings that impact the market where the modern ban on media advertising has had a significant impact.

Indigenous peoples in western society suppress that fact in space with ice-like movements, as they have emerged as consumers. In addition to the extra buy button, social media is already on hand - 20% of Facebook website visitors for all human beings and generations use their market every month, entirely from furniture to sports activities consoles. Platforms like Instagram inspire customers to take you as a market and store. Smaller groups can locate international shoppers with the aid of maintaining solely online (González-Padilla & Tortolero-Blanco, 2020). Success right here can exchange the enterprise model.

The epidemic has compelled many companies of all sizes on the verge of giving way that solely complies with ordinary buying and selling methods. Navigation light is also essential, as is geo identification - which capability altering one-of-a-kind content for managers relying on their current location. Snapchat and TikTok are complicated to ignore. The culture of touch that a publication the new generation affirms its importance. Snapchat has accelerated its unpopular provision for taxpayers to be the first to use the test experience. Primary producers have moved to TikTok, a bank with the promise of a network. Influencers sense strongly that they influence nearby markets, which is essential for products that strive to get customers all their facts at low-cost prices.

INDIA'S DIPLOMACY IN OPPOSITION TO COVID-19

India's Speedy Response

The world has closely monitored the spread of the coronavirus in India and India's response mechanism. Under the leadership of Prime Minister Shri Narendra Modi, India is fighting this virus with all its might. India is under complete lockdown for 21 days starting March 25 under the Natural Disasters Act, 2015. The initial announcement of the key occurs when the number of infected people is less than 400 once the WHO receives it. After that, the pressure on COVID-19 was created, the order for "social distance" and some extreme measures was announced. These necessary steps are described below.

- Contact tracing has started for people affected by the use of COVID.
- All visas are suspended
- All international and domestic flights, trains, and buses have been suspended.
- He initiated severe financial measures during this period to ensure no one went hungry.
- Indian railways converted carriages into isolation units.

Institutions Taking Up The Challenge

India's proactive, preventive, and "whole-of-government" approach to combating the COVID-19 pandemic is slowing down. The pandemic is ongoing, and slowing down exchanges between India and the rest of the world is counterproductive. This slowing of change disrupts the supply chains of many essential commodities required for combat. This slowing of change disrupts the supply chains of many essential commodities required for combat. The list of these vital commodities includes

- COVID-19 test kits,
- Masks,
- Alcohol-based disinfectants,
- Personal defensive equipment (PPE),
- Apparel materials for frontline fitness workers,
- Ventilators for patients.

India's Company and Its Efforts

The Department of Science and Technology (DST) is India's premier science and technology agency. With the help of institutions under the DST and sister ministries, the DST is taking the lead in coordinating efforts to map out and improve India's unique technologies to address the many challenges associated with COVID-19. DST, through its autonomous institutions and statutory bodies, has implemented three ways to fight against COVID-19:

- Huge opportunity mapping requiring R&D support, startups with viable products requiring production facilitation and support.
- Identification of market products requiring seed support.
- Help existing solutions that require significant growth to expand their manufacturing infrastructure and capabilities.

Research Proposals Invited By TDB

The Technology Development Board (TDB) set up under the DST has invited proposals from Indian groups and institutions to address protective measures and home care for patients suffering from COVID-19. Advanced technology or imported revolutionary solutions can assist, such as low-cost masks, thermal scanners, extensive sanitary and contactless access technology, rapid diagnostic equipment, oxygen, and ventilators.

Artificial Guide Breathing Unit (AMBU)

Sri Chitra Tirunal Institute of Medical Science and Technology (SCTIMST), Trivandrum, has developed an airflow device based on the Artificial Manual Breathing Unit (AMBU). The institute's AMBU automatic ventilator, with input from clinical faculty, will help critically ill patients breathe without the need to enter the ICU ventilator. The research is soon progressing to clinical trials and manufacturing through Wipro3D, Bangalore. In addition to these emergency airlifts, the agency is also trying to develop low-cost digital X-ray detectors with artificial intelligence to screen people suffering from COVID-19.

Antimicrobial Coating

Jawaharlal Nehru Center for Advanced Scientific Research (JNCASR), an autonomous arm of DST, has developed a one-step curable antimicrobial coating. This coating can kill influenza viruses, methicillin-resistant Staphylococcus aureus, fluconazole-resistant influenza viruses, and resistant pathogenic microorganisms and the severe acute respiratory syndrome coronavirus 2 (SARS-COV-19) virus series. The coating prevents microorganisms from becoming viable on the coated surface. During the COVID-19 outbreak, this layer can protect personal protective equipment, clothing, and healthcare equipment.

General Innovation

The National Innovation Foundation (NSF), like the self-supporting DST organization, encourages and supports important innovations developed by individuals and communities in the field of technology. NSF invites citizens to come up with creative and innovative ideas through the COVID-19 Challenge Competition (C3) to address following challenges:

- Nutrient Benefits and Immunity.
- Reduce the spread of the coronavirus.
- Disinfection of hands, body, household and internal equipment.
- Provide and distribute virtual goods to people, incredibly lonely older adults.
- Achieve people's relaxation at home.
- PPE and fitness equipment rapid diagnostic equipment.
- Reassess the "uncontacted" section for post-corona implementation needs and clear preferences of the population segment during COVID-19.

This initiative will encourage active participation in the government's anti-pandemic program and build scientific character among the public.

IMPACT EFFECTS OF COVID-19 ON SOCIETY

- Social media will continue to be relevant. Internet use has increased; when the presence expires, consumption must be measured. The influence of communication will be viewed no longer only in e-Commerce.
- Involvement with personal channels is increasing. Successful product availability on these channels is essential.
- As customers regularly do no longer favor packing public spaces, live streaming provides visible encounters. Civil society organizations and science groups prioritize the improvement and growth of equipment to aid the exercise if it proves to be a leading broadcaster.
- TikTok and Twitch create less expensive journey information with high-impact links. They are warm spots for creators and influencers. They need top content material products to supply content that meets activity organizations and a listing of younger people.
- Influence, message, and platform can assist form the company itself as an excellent advertising alternative for these channels.
- Never earlier than have human files been in a position to talk so rapidly in the face of an epidemic, social media has become increasingly needed to disseminate information; however, there are many blessings and

risks to consider. The use of this pc equipment can help speed up the spread of new information, new scientific discoveries, the distribution of diagnostic, therapeutic, and tracking purposes, and the trying out of a range of worldwide methods, doing away with national borders for the first time in history.

- To use these tools efficiently and efficiently, you are welcome to check out easy recommendations when sharing statistics on social networks at all times on COVID-19. We summarize the relevant statistics on the implications, benefits, and dangers of using social networking websites during the COVID-19 epidemic.

- Social networking structures are among the most broadly used statistics sources globally, easy and low priced web get admission to and the massive wide variety of users registered on these programs make them one of the satisfactory approaches to existing information. During main events, every response usually requires accurate records of whether or not the match is a tournament or a disease, or a natural catastrophe or not.

- The outstanding vogue can be viewed in the excessive quantity of Internet and telecommunications searches in China preceding severe COVID-19 instances in 10 to 14 days, the place Internet searches and social media are linked to outbreaks.

- Social media structures and being beneficial to the standard public to contact pals and family minimize the loneliness and loneliness associated with long-term nervousness and grief. Hence, it was essential to isolate domestic isolation to limit the psychological impact.

- One of the finest communication tools in the epidemic has been the rapid distribution of regional, national and global agreements. Sharing medical protocols, non-governmental resources, or suitable distribution indicators in needy clinical aid settings is now mainly new.

- This approves low-power establishments to implement high-speed strategies to operate or adapt different contracts with their use or offerings in the brief term, something unthinkable 20 years ago when many verbal exchange structures were unborn. We have supplied this manuscript, the benefits and issues associated with the use of social media during the epidemic.

Benefits of Latest Trends Using Social Media on Covid-19

- Social media has an outstanding benefit over the speedy distribution of analyzing content material in COVID-19 times, for example, to decorate the descriptive photo of air visitors who control suspected or licensed patients with COVID-19. It used to be shared on Twitter and WeChat; within a few days, it was requested to be translated into more than a dozen languages. Otherwise, the ado was allowed to convert files to a variety of health care facilities (Radwan & Radwan, 2020).
- The speedy distribution of data with waterproof ability has many advantages. Current findings involving the review of the hundred most seen videos on YouTube beneath the name "novel coronavirus," blended have more than 165 million views as of March 5, 2020, 85% of which are facts channels. It has been observed that less than ⅓ of videos talk about recommended recommendations, less than 1/2 talk about the most common symptoms; however, about 90% commented on death, anxiety, and racial segregation (Cinelli et al., 2020). This learning also leaves us with a necessary picture of the achievable for poorly transmitted transmission of suitable archives as properly as frequent symbols and symbols of COVID-19 in programs such as YouTube, which are widely mentioned as information.
- When publishing, observations have validated that the distribution of scientific literature on social media (Facebook, Twitter, etc.) will extend downloads. Several questions and complaints about COVID-19 are rapid national warning signs for sharing information, extra and more often. Specially modified editing, from months of adjustment to days or weeks viewed for adoption

Guidelines for the most reliable use of verbal exchange in information dissemination

1- Select distributions using embedded technology platforms or exchange groups.
2- Provide provision when sharing information. Avoid sharing statistics without a precise and dependable source.
3- Avoid sharing data that can purpose panic or anxiety.

4-   Quality is chosen over most of the place data allocation; in vitro studies and low-level evidence do not help daily work and can furnish baseless hope.

5-   Announce combat of interest, the place appropriate.

6-   Avoid giving scientific advice on social media and cease giving unsubstantiated references, which can confuse the installed community.

7-   Use precise peer evaluation and commenting methods, such as post-published peer evaluations or pre-print (unpublished manuscripts) such as medRxiv.org, furnish creator/institutional communication, and pursue the peer evaluation method as soon as possible.

- ○   Another benefit of the prevalent communication structures in the COVID-19 epidemic has been the opportunity to arrange collaborative research projects, research, and multidisciplinary studies. Finally, every other gain of social media systems that help similarly scientific training with online webinars recorded on applications such as YouTube, Skype, or Zoom.

Various Digital Technologies are Used to Overcome the Covid-19 Pandemic Effect

According to the WHO, the best way to reduce the spread of disease is to use it at an early age. With a strong focus on fitness care packaging, to deal with the epidemic is the use of intensive care and high-quality treatment for people living with HIV. In addition, reducing infection and bringing about tolerance in today's epidemic is one of the best ways for governments to work together to close the affected areas. Therefore, a secure community-building framework is needed to meet the needs of a health machine and use it to shut down the work (Atlani-Duault et al., 2020). Under these circumstances, applied science has impacted various sectors like health care, government, society, industry, etc. The sudden switch to science has accelerated the search for new and innovative products and has forced technological barriers in almost every field. Handling the burden of previous military forces, technologies such as telehealth, artificial intelligence (AI), robots, drones, etc., have multiplied many times over and over again in this epidemic. Drone and robot technology know-how is designed to assist in disinfection, transport clinical material, surveillance, trying out of humans for early detection of the virus, etc. In addition, robots have been widely used to provide scientific and in-depth consultations. The fast enlarges in COVID-19 stipulations and the inclusion of human security has created the want to analyze large numbers to make agreements to address the disease (Khan et al., 2021). With the capacity to work independently, do a lookup to enhance via adding new important points and drawing conclusions in the brief term. AI has been used to predict regions that threaten pollution, drugs, laptop diagnostic results, etc. Or the mobile community is trusted as one of the key ways to deal with the disease. In the past, mobile technology knowledge has emerged in every city and connects people to large crowds. Seeing the mobile network as a stream of smartphone apps (Applications) has been an excellent template for joining more people (Sandre, 2020). The various cellular applications are designed to express community focus, provide clinical assistance, and, most importantly, stop the emergence of COVID-19 through a series of contacts. As we come together to operate pursuits duties daily, the authorities are focused on preventive and emergency measures. On the different hand, researchers worldwide are getting involved in a growing care market that will help the speedy development of COVID-19. Combined efforts from a few sectors, technological implementation, and cooperative governance have created the COVID-19 defense line. Its remaining evaluation highlights the hidden roles of intelligent science and exhibits its ability to deal with epidemics. These findings incorporate a one-of-a-kind investigation of effective technology-based strategies that contribute to epidemic administration and the evolution of advanced science to the great of its ability. This report helps academics, professionals, and students see how their use of technology helps manipulate the prevalence of novel coronavirus and, in turn, promote the use of this science in modern or future emergencies. Drone Science Authorities use drones for fitness care, surveillance, disinfection, infection checking out, and public awareness. Drone's technological know-how can perform all of these features and permits its operation in internal areas without human interaction (Bjola & Manor, 2020).

Delivery

In novel coronavirus infection, drones are widely familiar for the rapid and secure transport of simple products. The transport method can be divided into two categories: the provision of meals / primary necessities and the transport of fitness care services, including laboratory samples, labs, community safety gear (PPE) kit, etc. In addition, the donation device primarily based on donations does not contain humans in this system which helps minimize the unfold of novel coronavirus and decrease shipping time. Monitoring and broadcasting in many nations took steps such as closing public places, suspending public meetings, and the policy of social segregation to stop physical contact. There is an exceptional debate over how governments, organizations, and societies work to preserve little effort except for bodily contact. In this case, authorities use drones to expose the pastime and avoid social gatherings that should be detrimental to the community. Drones used in broadcasting and public broadcasting with speakers and even digital signage are broadly used in many countries. Using a world positioning gadget (GPS) enabled by utilizing the drone administration system, Authorities can quickly locate an unpopular job even in densely populated areas (Moumita de das, 2020).

Antibiotics

Handling novel coronavirus contamination with preventive measures is sluggish and can take months; therefore, governments have furnished an answer because of its closure to stabilize the economy. As we collectively go out and raise the norm in day-by-day life, we must kill germs often at excessive altitudes.

Authorities have come up with an agricultural answer for spraying drones so that germs can be killed in an unsafe place every second. These drones' blanketed region and speed can now be uneven compared to the exclusive spraying methods.

Testing

Human contact is one of the common factors of transmission of novel coronavirus. Therefore, it is vital to pick out the contaminated personality and the individual who may have been contaminated due to early contact. The concept is multidisciplinary behavior and speedy research surrounding areas with little personal connection. To improve the correctness of the sample test, the technology that tests faraway respiratory stipulations is similar to thinking about drones and is acknowledged as the "Pandemic Drone." Initially, epidemic drones have been designed to observe drug disasters and struggle areas remotely. Various technologies encompass a computer monitoring machine and a one-of-a-kind sensor to detect breathing levels, heart rate, temperature, and cough in crowds. To combat the world presence of the novel coronavirus in the USA government have developed a virus called 'Dragonfly' that video display units and detects human beings with infectious and respiratory infections. Science provides energy trying out in a massive location in a short period, and an enormous range of human beings can be seen in many conditions and daily.

Robotics Technology

Robotic technology has confirmed its greatness in a wide area, which can be helpful in many ways. Robotics technology has become increasingly influential in many areas such as drug delivery, surveillance, an affected person trying out and disinfection, etc. For example, robots are used in bulk checks and locate facts to interpret each health document's reviews (Grincheva, 2021). In epidemic cases such as the novel coronavirus, the robotic device has provided sizable help to health employees via performing low-frequency repetitive tasks. Robot technology can be exclusive mainly through the first-rate degree of conversation between robots and humans.

Telerobot

Tele-robot is a science that can be used to manage robots away from a Wi-Fi communication neighborhood such as the Internet. In the aftermath of the COVID-19 epidemic, artificial robotic exercising machines proved the most secure and environment-friendly technology. Robots can also be produced besides challenge from time to time, making them briefly tremendous for trying and feeding patients. The 'intuitive' developed 'DaVinci Surgical Robot' has received much attention in this discipline. It operates with a surgical software application that can be used with the assistance of a physician. Another instance is using the arm of robots working on telephones to perform ultrasound tests.

In addition, robots are used in hospitals and neighborhoods to perform secondary duties such as testing, rehabilitating patients, and bringing remedy all through recovery. Tele-robotics that help rehabilitate the fence are used to speak to the sufferers who stay in it and inform them of their cognitive or bodily rehabilitation activities.

Independent Robot

Independent robots are active in making unbiased preferences and can take significant steps to stop human intervention. During COVID-19, robots are very environment friendly and are widely distributed in UV reproductive robots. A robotic with UV-disinfectant wheels are used in hospitals, airports, grocery stores, shops, etc. Using robotics technology with fraudsters can continue to work on unique treatments. An independent robotic is used to supply PPE, medicine, tools, and food and prepare preparatory work.

Wearable Robot

The science of robotics separately refers to technological gadgets that can be worn on the body and have the energy to make and make selections while helping human beings in computer design, trying to communicate. One such instance is warm shoes in areas dealing with a massive variety of roads every day. Recently, a helmet has been used to monitor temperatures and recognize facial aspects to show a pedestrian's small print.

Artificial Intelligence

Artificial Intelligence has been viewed as a possibility that arises on the Internet and in the world itself. Many nations are using intelligent technologies to expose novel coronavirus infections, inform fitness care systems, and devise ways to curb their spread.

Covid-19 AI-Based Modern Strategies

The world is exploring powerful technology areas, specifically AI and mathematical science, to test and fight the epidemic. More recently, Australia's census has centered on censuses (NHSX, 2020).

The Epidemic Model, earlier than it used to mimic the flu epidemic that was once used to modify and be measured to mimic the COVID-19 epidemic all through Australia. China has adopted an in-depth learn about mannequins based on authentic COVID-19 studies in polluted regions. Various data have been compiled, such as the death toll, the spread of social media, overcrowding, and demographics, to inspect the threat of contamination in the region (Lancet, 2020). COVID-19 contamination testing framework in the structure of a touch display smartphone constructed into AI technology. Throughout the ultimate work, in-depth practice on diagnosing document acquisition and job evaluation is related to ailment symptoms. For example, films and pics purchased on a smartphone digicam or sensory tests are no longer a lesson to analyze human fatigue tiers (Jayatilaka, 2020).

Similarly, the cough situation can be decided from the sound recording on the smartphone microphone. Or, a video uploaded to a smartphone can assist in predicting nausea. The company will focus on neutral programs, robots, AI, and blockchain science to deal with robust systems. In the same book, health analytics and the'Cotiviti'' response agency now use AI and vital records to predict the warm spots of the novel coronavirus around the USA. Many applied scientists have long gone via a developmental method that will help deal with vaccines such as the novel coronavirus (Purwasito & Kartinawati, 2020).

AI drug-based novel coronavirus, and mathematical calculations are essential for epidemic conditions such as COVID 19. Post findings to assist others in improving drug resolution and better become aware of the virus. AI has been using AI fashion to make capsules that can remedy different serious ailments and make an effort to deal with COVID-19 (George & Paul, 2020). Using intelligent technologies sometime in the first few weeks of the epidemic itself, it has counseled cutting-edge pills to help treat novel coronavirus infections. Large-scale equipment and cloud gear worldwide are thinking about possible solutions for the novel coronavirus vaccine. These frameworks can run the calculation of fashions and options at a faster fee than known laptop processing. AI to examine and diagnose new tablets with excessive anti-novel coronavirus effects. Ongoing efforts are underway to construct cost-efficient science checks with SARS-CoV-2.

Epidemic to Phone Use

Since the outbreak of the COVID-19 epidemic, a wide variety of smartphones have been developed, some working with authority'' officers and others working with personal companies. Global governments and retailers remember smartphone apps to grant brand new coronavirus information, notice dementia, supply fitness services, and assist diagnose you. Depending on the type of activity, cell functions can be categorized as Virtual Apps, Self-Testing Programs, and Contacts Tracking programs.

Notification Apps

These things to do now do not require much work as they are designed to provide facts differently from communication. These packages supply current COVID-19-related data such as pinnacle stories, truth sheets, guidelines, etc.

Self-Assessment Programs

COVID-19 phone assist programs to assist humans in testing their fitness and getting instructions/references from COVID-19. The purpose is to furnish a certain level of grasp to people who exhibit signs and symptoms of contamination and are now not accustomed to being unwell with easy questions. In addition, it serves as a tool to instruct human beings on how to get the sickness and information them through preventive measures.

Contact Monitoring Apps

Monitoring and warning are essential functions in all stages of an outbreak. The significant objectives of tracking and warning are to become aware of the suspect and ask them to isolate themselves if they exhibit signs. Communication monitoring helps assemble documents to perceive methods to spread air pollution in the community. The number of purposes for monitoring contacts follows applied sciences such as Bluetooth specifications, cryptography specifications, and an interface program (API) framework.

X-Ray and CT Imaging Technology

The COVID-19 neural (COVNet) community is an in-depth knowledge of algorithm that scans CT photographs to realize novel coronavirus in patients. The non-stop detection of the framework is based totally on the integration of three-d CNN ResNet-18 neighborhood and website online monitoring proposed to analyze pulmonary CT pictures to observe COVID-19 infection. The infection of Noval coronavirus Chest CT scan Evaluation System hooked up in Shanghai should notice suspected suspects within seconds according to the clinical expansion of COVID-19 the use of a comprehensive screening method to screen CT-scan snap pictures with 89.5% accuracy. CT scans of people infected with the novel coronavirus were analyzed to compare the demise of an infected personality.

CONCLUSION

COVID-19's emergence has pushed the boundaries of technology and accelerated innovation in every industry. Technology's assistance in managing the pandemic crisis is invaluable. Modifications to existing drone, robotics, and artificial intelligence technology and the context in which they are used to battle pandemics have been thoroughly studied. The findings show how innovative technology-based solutions that connect vast people may be set up to deal with emergency circumstances. The analysis delves into new advancements in the latest sensor technology that have been examined around the world during the pandemic. A more comprehensive picture of how various technologies are applied and the modifications they have experienced worldwide to directly or indirectly aid in the pandemic crisis is offered. The impact of these technologies on society in numerous aspects has been summarized. As a result, the Indian discovery aids in comprehending the influence of technology on society and how these advances might be fine-tuned for emergency response in the future.

REFERENCES

Abbas, J., Wang, D., Su, Z., & Ziapour, A. (2021). The Role of Social Media in the Advent of COVID-19 Pandemic: Crisis Management, Mental Health Challenges and Implications. *Risk Management and Healthcare Policy*, *14*, 1917–1932. doi:10.2147/RMHP.S284313 PMID:34012304

Ahmad, A. R., & Murad, H. R. (2020). The Impact of Social Media on Panic During the COVID-19 Pandemic in Iraqi Kurdistan: Online Questionnaire Study. *Journal of Medical Internet Research*, *22*(5), e19556. doi:10.2196/19556 PMID:32369026

Atlani-Duault, L., Ward, J., Roy, M., Morin, C., & Wilson, A. (2020). Tracking online heroisation and blame in epidemics. *The Lancet. Public Health*, *5*(3), e137–e138. doi:10.1016/S2468-2667(20)30033-5 PMID:32085818

Bjola, C., & Manor, I. (2020). *Digital diplomacy in the time of the coronavirus pandemic.* USC Public Diplomacy. https://uscpublicdiplomacy.org/blog/digital-diplomacy-time-coronavirus- pandemic

Cinelli, M., Quattrociocchi, W., Galeazzi, A., Valensise, C. M., Brugnoli, E., Schmidt, A. L., Zola, P., Zollo, F., & Scala, A. (2020). The COVID-19 social media infodemic. *Scientific Reports*, *10*(1), 16598. doi:10.103841598-020-73510-5 PMID:33024152

Depoux, A., Martin, S., Karafillakis, E., Preet, R., Wilder-Smith, A., & Larson, H. (2020). The pandemic of social media panic travels faster than the COVID-19 outbreak. *Journal of Travel Medicine*, *27*(3), taaa031. doi:10.1093/jtm/taaa031 PMID:32125413

George, B. & Paul, J. (2020). *Digital Transformation in Business and Society Theory and Cases: Theory and Cases*. Springer. . doi:10.1007/978-3-030-08277-2

González-Padilla, D. A., & Tortolero-Blanco, L. (2020). *Social media influence in the COVID-19 Pandemic*. National Institutes of Health., doi:10.1590/S1677-5538.IBJU.2020.S121

Grincheva, N. (2021). Digital diplomacy in the midst of (post) pandemic crisis: Inspiring, educating, and contributing to global peace and well-being. *MW21*. https://mw21.museweb.net/paper/digital-diplomacy-in-the-midst-of-post-pandemic-crisis-inspiring-educating-contributing-to-global-peace-and-well-being/

Jayatilaka, C. (2020) The Effects of Digital Diplomacy on International Relations: Lessons for Sri Lanka https://lki.lk/publication/the-effects-of-digital-diplomacy-on-international-relations-a-lesson-for-sri-lanka/

Jiang, Y. (2021). Problematic Social Media Usage and Anxiety Among University Students During the COVID-19 Pandemic: The Mediating Role of Psychological Capital and the Moderating Role of Academic Burnout. *Frontiers in Psychology*, *12*, 612007. doi:10.3389/fpsyg.2021.612007 PMID:33613391

Kadam, A., & Atre, S. (2020). Social media panic and COVID-19 in India. *Journal of Travel Medicine*, *27*. doi:10.1093/jtm/taaa057 PMID:32307545

Khan, H., Kushwah, K.K., & Singh, S. (2021). Smart technologies driven approaches to tackle COVID-19 pandemic: a review. *Biotech*, *3*(11), 50. doi:10.1007/s13205-020-02581-y

Lancet. (2020). COVID-19: Fighting panic with information. *Lancet*, *395*(10224), 537. doi:10.1016/S0140-6736(20)30379-2 PMID:32087777

Madan Gopal, K. (2020) Importance Of Civil Society Organisations In Managing Covid-19 Pandemic. *Outlook India*. https://www.outlookindia.com/website/story/india-news-the-importance-of-the-role-of-civil-society-organisations-in-managing-the-pandemic/364223

Manor, I. (2021). The Digital Legacy of Covid-19 https://www.e-ir. info/2021/04/02/the-digital-legacy-of-covid-19/

Moumita de das. (2020). *Digital diplomacy: the trend is here to stay.* Adamas University. HTTPS://adamasuniversity.ac.in/digital-diplomacy-the-trend-is-here-to-stay/

Mullard, S., & Aarvik, P. (2020). Supporting civil society during the Covid-19 pandemic. The potentials of online collaborations for social accountability. *E-International Relations.* https://www.u4.no/publications/supporting-civil-society-during-the-covid-19-pandemic

NHSX. (2020). *Driving forward the digital transformation of health and social care.* NHSX. https://www.nhsx.nhs.uk/.

Purwasito, A., & Kartinawati, E. (2020). *Hybrid Space and Digital Diplomacy in Global Pandemic Covid, 19.* doi:10.2991/assehr.k.201219.100

Radwan, E., & Radwan, A. (2020). The-spread-of-the-pandemic-of-social-media-panic-during-the-covid-19-outbreak. *European Journal of Environment and Public Health, 4*(2), em0044. doi:10.29333/ejeph/8277

Rick, J. (2020). *3 reasons why civil society is essential to COVID-19 recovery.* WEF. https://www.weforum.org/agenda/2020/05/why-civil-society-is-essential-to-covid-19-pandemic-recovery/

Sandre, A. (2020). In review: top 10 moments in digital diplomacy. *The Medium.* https://medium.com/digital-diplomacy/2020-in-review-top-10-moments-in-digital-diplomacy-57b802e0159c

Shankari, S., Rani, L., Brundha, & Somasundaram, J. (2020). Knowledge and Awareness on Role of Social Media in Managing COVID-19 Among General Population- A Questionnaire Study. *International Journal of Current Research and Review, 12*(19), 197–202. doi:10.31782/IJCRR.2020.SP25

Turchetti, S., & Lalli, R. (2020). Envisioning a "science diplomacy 2.0": On data, global challenges, and multi-layered networks. *Humanit Soc Sci Commun, 7*(1), 144. doi:10.105741599-020-00636-2

Volodenkov, S., & Pastarmadzhieva, D. (2020). Digital society in the context of the COVID-19 pandemic: First results and prospects (comparative analysis of the experience of Russia and Bulgaria). *Journal of Political Research., 4*(2), 80–89. doi:10.12737/2587-6295-2020-80-89

Chapter 8

Dehumanising and De-Africanising Public Diplomacy:
A Philosophico-Cultural Perspective on the Digitalisation of African Diplomacy

Floribert Patrick C. Endong
https://orcid.org/0000-0003-1893-3653
University of Dschang, Cameroon

ABSTRACT

The COVID-19 pandemic brought to the fore not only the centrality of Western digital technologies, but also a number of philosophico-cultural issues. Two of such cultural issues in the domain of digital diplomacy have been the de-Africanisation and dehumanisation of African digital diplomacy. These two issues have partly stemmed from the popular African myth that cybercultures in general and the digitalisation of African diplomacy in particular are disruptive forces that could negatively affect the traditional African values that have since independence been upheld by African governments in their conduct of public diplomacy. A related theory states that digitalisation may only de-humanise and de-Africanise public diplomacy. Using secondary sources and critical observations, this chapter examines the extent to which the above-mentioned fears are justified. The chapter specifically explores how digitalisation could affect specific African traditional values. It also examines the extent to which digitisation is susceptible to de-humanise or/ and de-Africanise African public diplomacy.

DOI: 10.4018/978-1-7998-8394-4.ch008

INTRODUCTION

Many schools of thought have construed digital diplomacy essentially as a technological revolution, giving little or no attention to the human or socio-cultural dimensions of the phenomenon. It is, in effect, common to come across both African and non-African critics who merely regard e-diplomacy as the "banal" application of ICTs in the conduct of public diplomacy. Viewed from this standpoint, digital diplomacy is most often construed as a technology-driven paradigm which only or mainly facilitates old behavioural models in public diplomacy or again, as a force which disrupts analogue practices in public diplomacy but still leads to the same finality as in "non-digital" or "classical" diplomacy. Besides these observations, there are sceptical critiques which even totally negate the influences and potential longevity of digital diplomacy. A case in point is Former US Secretary of State John Kerry who in 2012 sounded as though, digital diplomacy is a not-too-new concept and a phenomenon which does not deserve a special consideration. In a reaction which many critics viewed as hastened, Kerry claimed that "the term digital diplomacy is almost redundant - it's just diplomacy, period" (cited in Bjola 2018). Similar sceptical and reductive perceptions of digital diplomacy were expressed by some Kenyan diplomats in a more recent survey conducted by Waithaka (2018). The survey concretely revealed that a number of staffers at Kenyan MFAs categorically disregard the digital diplomacy concept, viewing it simplistically as "a passing wind that will, [in] no way change the conduct of diplomacy" (Waithaka 2018: 113).

Meanwhile, the influences of the digital technologies on the practice of public diplomacy are not only real; but go beyond the simple transition from one technology to the other (analogue to digital). Indeed, the influences entail a deal of revolution in diplomats' world views, systems of values and ideology. This is so as diplomats are first and foremost social beings, in the same way foreign ministries are social institutions. This also follows the logic that digital technologies definitely influence the norms, values and working philosophies of diplomatic institutions. No doubt, terms such as "diplomatic norms", and "norms of diplomatic culture" have since become aphorisms in the public diplomacy sphere. Thus, as a process and paradigm which impacts the African society, digital diplomacy is bound to affect African public diplomacy not only from the technological, but also ideological and cultural points of view. In other words, digitalisation is likely to affect some of the socio-cultural values African governments have upheld since independence in their conduct of public diplomacy.

Studying the extent to which specific cultural norms of African diplomacy could be affected by digitalisation or other post-modern technological innovations has not really attracted the attention of both local and exogenous researchers; thus making the concern of this paper topical. In view of filling the above mentioned apparent gap in knowledge, the present chapter specifically explores the possible effects of digitalisation on such African traditional values as (i) seamless approach to the passage of time, (ii) respect for cultural tradition and authority, (iii) predilection for the collective, (iv) unhurried decisions, and (v) the prioritisation of community rather than individuals – aphorisms believed by many scholars (notably Spies 2018) to have governed African public diplomacy for decades.

The chapter also sets out to examine the extent to which digitalisation could dehumanise and de-Africanise the conduct of public diplomacy in the African continent. Going by the two above objectives, the paper answers the following research questions: How have African traditional values been reflected in Africa diplomacy? How is digitalisation a threat to, or an opportunity for the application of African traditional values in the conduct of African diplomacy? How justified is the myth that digitalization is a threat to the upholding of African traditional values in the conduct public diplomacy? And how is the digitalisation of African diplomacy compatible with the concept of African humanism?

CONCEPTUAL AND THEORETICAL ISSUES

It will be expedient from the outset to provide the conceptual definition of two key terms/theories that will be used in this discourse namely dehumanisation and de-Africanisation.

Dehumanisation

Authors have defined dehumanisation as (i) the act of subjecting others to indignities or the Kantian concept of treating others as a mere means, (ii) a rhetorical act in which a human being is metaphorically likened to a non-human entity (an animal or an inanimate object); (iii) the denial of subjectivity, individuality, agency or distinctively human attributes to others, (iv) the act of viewing others as sub-humans and (v) the act of treating others in a way as to erode, obstruct, or extinguish some of their distinctively human attitudes (Livingstone 2011; Haslam 2006; Masolo 2018; Zaharna 2018). All the above definitions indicate that dehumanisation is, in principle, concerned with the "maltreatment" of human beings and not things or abstract concepts. However, in this paper, dehumanization will be seen as the removal or reduction of human involvement or interaction in the practice of public diplomacy. This is in line with the arguable but pertinent notion that all human activities are in the best of situations, humanised.

The dehumanisation of a human industry or enterprise is often characterised by the systematisation, homogenisation, and mechanisation of human work. According to Baker (2016) it definitely entails subjecting workers to performance management rituals that tend to remove or downplay their humanness. It also means expecting the workers to religiously follow instructions; to not upset the apple and to stick to the straight and narrow, just like machines. In addition to these, Baker contends that dehumanisation implies referring to workers merely as "human capital" and "human resources", and thus turning them to abstract pieces of machinery production. In brief, the dehumanisation of work or human industry entails turning a human activity to a cold, clinical, rational domain devoid of humanness.

One major characteristic of the dehumanisation of human activities rests in the use of automation and the progressive replacement of human agency with machines or machine-assisted processes. This is today frequently seen in the growing use of machine intelligence and conversational experiences in key human industries such as public health, transportation, courts, education and in business enterprises among others. These machine-driven processes most often create near-human interactions that paradoxically reduce the much-needed human touch in many workplaces and human industries. The application of the digital technologies and robots in specific human industries has in some ways, reduced human engagement and caused human touch in these industries to likewise be reduced or removed totally. With close reference to the management of human resources in many contemporary workplaces, the online magazine *HR Technologist* (2017) observes that human industries' constant and growing resort to Human Resources Technologies and automation as a cost-effective tool for production is indicative of the fact that in a near future, work will entail that man and machine work side by side. With transactional and mundane jobs increasingly being given to machines and with the introduction of such technologies as Chatbox in businesses and organisations to replace employee-employers conversation, humanity is progressively driving to a future where most human activities are dehumanised.

In line with the above, there have been fears that digitalisation and automation are dehumanising – or have the potential to dehumanise – the conduct of public diplomacy. These fears rest on at least two factors. The first factor revolves around the popular myth that digital technologies such as the social media have dehumanised (or at least denatured) most interactions between diplomats and between diplomats and foreign audiences, by taking them (the human interactions) to a "nowhere called cyberspace" and by reducing or removing humanising aspects of typical interactions such as the sense of touch, smell and non-verbal cues (Hymowitz 2018; Shwad 2019). The second factor is rooted in the observation that the dark sides of the internet have facilitated the emergence of robot-human interactions on the cyberspace. In other words, innovation in the domains of computation and AI has bred scenarios wherein diplomats are often made to interact with machines (robots) rather than with human beings. In effect automation and computation have made social media actors not necessarily to be humans. Today it is even becoming common to find "bots" operated by a single programmer which mass produce digital contents, flood the social media system with all manner of false, misleading, and anachronistic messages and seek to interact with foreign audiences and even diplomats, embassies and MFAs in an unhealthy manner (Unver 2017). Besides this, the mass production of digital diplomatic messages by "bots", advances in automation, robotics and computations are all making the concept of robot diplomat a new paradigm in public diplomacy. Kurbalijah (2020) for instance reviews cases and some thought-provoking evidence of uses of AI in international diplomacy. He underscores China's constant leverage of the AI technology to provide insights for its diplomats into the possible scenarios and the evolution of events on the international arena. Kurbalijah also explores the use of AI in supporting economic diplomacy in trade negotiations.

All the evidence mentioned above and many other similar robot-driven paradigms – that will be explored later in this essay – have made automation not only to sit at the centre of modern diplomacy but to also strongly contribute in dehumanising modern public diplomacy. This dehumanisation will be treated in greater details and vis-à-vis the concept of African humanism and other selected African cultural values in the upcoming sections of this discourse.

De-Africanisation

The term de-africanisation could be said to come from the verb "Africanise." To Africanise means at least three things: (1) to replace the non-African features or components of a concept with Black African ones; (2) to bring something under the influence of Africa(ns) and (3) to adapt a concept to African needs (Lown 2010). In line with the second and third definitions mentioned above, the Sankofa Youth Movement construes Africanisation as a situation where people of African extraction embrace their African heritage and develop a sense of loyalty towards their continent. This culturally determined act entails adopting and promoting African cultural values, putting them on the pedestal currently occupied by the western values (cited in Louw, 2010: 42). Still in tandem with the definitions provided above, Louw (2010) contends that Africanisation reflects African's common legacy, history and postcolonial experience. Through this legacy, they have to connect with the broader African experience and establish concepts, processes, practices or paradigms that bind them together. They then "confront their own sense of Africanness, transcend their individual identity, seek their commonality and recognise and embrace their otherness" (p.42).

The above definitional illuminations suggests that the contrary term "de-Africanisation" is construable as a dramatic process whereby an originally African concept, idea or process is instead stripped of its key African cultural values. By the same logic, the de-Aficanisation of public diplomacy could be defined as a situation where the public diplomacy practiced by African MFAs and diplomats is stripped of its African character. To borrow Gluck's (2015: 2) language, the de-Africanisation of public diplomacy is "an *epistemic shift* away from ideas of parochialism [and Afrocentrism]", which have long defined African's MFAs and embassies' practice of public diplomacy.

In other words, the de-Africanisation of public diplomacy is a situation where a number of traditional African values stop being upheld by African MFAs and embassies in their conduct of public diplomacy. Defining these cultural norms and values in a clear-cut manner is a herculean task given the fact that the African continent is not monolithic but rather comprised of a mosaic of ethnicities, languages, and cultural groups. In spite of this monolithic definition of the African, a number of research works in history, philosophy, sociology, and anthropology has credibly highlighted cultural norms and values that are upheld across African communities or that are at least specific to the majority of African identities and nationalities. As noted by Idang (2015: 97), it may be faulty to presuppose that the totality of African societies has the same explanation(s) for events, the same language, and same mode of dressing and so on; "rather, there are underlying similarities shared by many African societies which, when contrasted with other cultures, reveal a wide gap of difference". In the same line of thought, Gordon (2020) contends that the existence of a plurality of African ethnicities does not cancel the fact that the same kind of cosmologies prevail across the African continent. These similar cosmologies ground Black African peoples' religious and cultural practices. Gordon further partially attributes the commonality of Africans ethnicities' cultural norms and values to their histories and geographical origins. He writes that many African ethno-religious groups are descended from a set of communities along the ancient lakes and plains of the Sahara-Sahelian region of northern Africa that subsequently dried up, becoming desert. This shared origin justifies similarities in their cosmologies. In guise of explanation, Gordon pointedly adds that:

The cosmologies of these groups [African various ethnicities] tend to have a concomitant ontology, or conception of being, and a system of values, in which greater reality and value are afforded to things of the past. Thus, the Creator, being first, has the greatest ontological weight, and whoever is brought into being closer in time to the moment of the origin of the world is afforded greater weight. This view gives one's ancestors greater ontological weight and value than their descendants. Also, one's past actions are of greater ontological weight than one's present actions. (One's future actions are of no ontological weight since they have not yet occurred.) Indigenous African systems affirm that human beings negotiate their affairs with the understanding that they cannot change the past (although they can be informed by it, especially through ancestors), are entirely responsible for the present, and must take responsibility for their future. This form of humanism does not require the rejection of religion, but may exist alongside it.

In line with the theory captured by Gordon above, a number of scholars – notably Spies (2018), Laverty (2015) and Murithi (2006) – have theorised that some of the cultural norms upheld by African countries in their conduct of public diplomacy include (i) seamless approach to the passage of time, (ii) respect for cultural tradition and authority, (iii) predilection for collective, (iv) unhurried decisions, and (v) the prioritization of community rather than individuals. These values will be treated in greater details in the following section of the discourse.

TRADITIONAL CULTURAL VALUES AND NORMS IN AFRICAN PUBLIC DIPLOMACY

Like any other human industry, public diplomacy is driven by well-defined norms, values, and ideologies. In his theory bordering on the diplomatic culture, Goefrey Wiseman (2005) contended that there exist five norms of diplomatic culture universally upheld by nations of the world. These include (i) the use of force as last resort, transparency, continuous dialogue, multilateralism, and civility. Although pertinent, this theory has been criticised on various grounds, two of which are the facts that, the theory mainly emanates from the Euro-centric Westphalian system and tends to overlook many peculiarities of the nations that are not part of the Anglo-Saxon sphere of geopolitics as well as countries that formerly belonged to the Non-Aligned block. Secondly, Wiseman's theory omits the norms of African countries among many other Global South countries. As noted by Laverty (2015), although they may eventually be accepted by African States, the five principles enunciated by Wiseman's theory, glaringly overlook the regional and cultural norms that may be specific to Africa. In guise of countering Wiseman's theory, one could therefore advance at least three to four norms governing African public diplomacy. These include (a) anti-imperialism and pan-Africanism, (b) the vehement adherence to the political borders left by the colonial powers and (c) solidarity of African states at world stage.

Indeed, anti-imperialism has, as a norm, driven African diplomacy immediately after independence. One reason for this resides in the fact that new forms of western political, economic and socio-cultural domination have emerged and remained perceptible in many if not the majority of Black African countries. In French speaking countries for instance, colonial currencies such as the CFA Franc are still prevailing; while colonial patterns of trade between France and her former colonies coupled with the military presence of such western and Asian countries as the US, France and China on various parts of the continent continue to be decried (Endong 2020; Transparency International 2019). This state of affairs has unsurprisingly pushed African leaders to continuously adopt an anti-imperial rhetoric in the design of their foreign policies and diplomacy.

In addition to this anti-imperialism rhetoric, the non-integrationists (African States not supportive of AU's clamour for political unity of the continent) have cherished the *uti possidetis* norm which revolves around the idea that no entity should tamper with the political borders left by the colonial powers.

Thus, many African countries have, in spite of their adherence to pan-African ideals, sought to stick to the lines drawn on the continent in guise of territorial borders by European colonial powers in Berlin in 1884-1885. The *uti possidetis* norm has in part been supported by the 1963 Charter of the Organisation of African Unity OAU which states that the regional organisation is:

Determined to safeguard and consolidate the hard-won independence as well as the sovereignty and territorial integrity of our states, and to fight against neo-colonialism in all its forms". The resolutions adopted at the first ordinary meeting of the OAU declared "that all Member States pledge themselves to respect the borders existing on their achievement of national independence.

The OAU has also backed the *uti possidetis* norm by showing reluctance to support secessionist movements on the continent. Egregious examples include the Biafra and MASOB movements in Nigeria, the Azawad movement in Mali, the Katanga in former Zaire (DR Congo), the Somaliland in Somalia and the recent Ambazonia separatist movement in Cameroon, which all never benefitted from the support of the OAU/AU. In these ways, the OAU/AU has strongly been supporting the national sovereignty of its member States. In tandem with this, the AU has always been working for the integrity of the colonial borders that define the territories of modern African countries.

Another norm governing African diplomacy is that of the solidarity of African States at world stage. This has played out in forums such as the UNO where African States have in many occasions, particularly on issues concerning their continent, spoken with one voice. Also, the perceived vulnerability of individual African nations has been an incentive for Africa's international relations to be expressed "collectively" through such processes as multilateralism. Spies (2018) even contends that postcolonial Africa's global diplomatic impact is attributable to its group efforts manifested through multilateralism rather than to individual African States. This tendency of speaking with one voice (collectivism through multilateralism) has, according to Spies, been both a practical advantage and "a matter of historical redress". The collectivism has also been a way through which African States have expressed their common cultural value of the predilection for the collective. Other cultural values upheld by them (African States) include (i) seamless approach to the passage of time, (ii) the respect for cultural tradition and authority, (iii) the unhurried decisions, and (iv) the prioritization of community rather than individuals.

The latter cultural values find expression in a number of popular concepts that highlight societal selfless and African communalism. Some of these concepts include the palaver tree (a forum widely used in traditional Africa to settled dispute), the *Harambee* philosophy (a Swahili born concept meaning "pulling together") and the famous *Ubuntu* (Nguni concept for "being human") among others. Many African governments have even anchored their diplomacy in the application of these societal selfless concepts/philosophies a good example being South African government's adoption of its White Paper of May 2011. This foreign policy document has as title "Building a Better World: The Diplomacy of *Ubuntu*". The Nguni Bantu term "Ubuntu", fuels the belief in a universal bond of sharing which connects all humanity. It specifically stems from the Nguni Bantu peoples' aphorism that the personhood of anyone depends on the community to which he belongs. Translated literally, "ubuntu" means "humanity towards others".

Another example of African countries which has based its diplomacy and foreign policy objectives on the predilection for the collective and the prioritisation of the community rather than the individual is Nigeria. This is seen the fact that the country's has based its foreign policy objectives on being "its brothers' keepers". As former director general of Nigerian Institute of International Affairs Gabriel Olusanya (2012: 27) puts it "Nature has so placed Nigeria in a situation where she has to be her brother's keeper. If your neighbours are hungry and unhappy, you can never be happy. Nigeria's neighbours are wretched and therefore we have to be our brother's keeper". The above reveals that African diplomacy has over the years been driven by specific norms and values. Examining the way the application of new technologies could affect their adoption by various African states is a topical question. This will be addressed in subsequent sections of this paper.

DIGITALISATION AND THE DEHUMANISATION OF AFRICAN DIPLOMACY

The advent of the Internet and the social media has led to the emergence and propagation of various technocultures on the African continent. One of these technocultures has been digital diplomacy which, like the plurality of other digitally driven paradigms, has not been spared by culturally and ontologically based criticism. An understudied aspect of this criticism revolves around the arguable myth stating that digital diplomacy is – progressively becoming – a dehumanised form of human enterprise. This myth has mainly been fuelled by speculations or theses which generically associate digital cultures as a whole with a subtle form of dehumanisation or a denaturalisation of human cultures. In an opinion paper published in the American tabloid *Wilmington Star News*, Hymowitz (2017: 18) contends that the social media dehumanise human interactions by taking them to an unnatural, immaterial, and unknown "nowhere" called the cyberspaces and by rendering them devoid of some characteristics specific to human communication such as non-verbal cues. Hymowitz additionally contends that social media communications also dehumanise human interactions by pushing the *internauts* to psychologically equate any human subject or human interest they discuss on social media with an abstract thing and not with a human. He illustrates his observation thus:

Social media dehumanises personal interactions, taking them out of the dining room, the neighborhood store and workplace and into a nowhere we call cyberspace. In real life, you're bound to run across some flesh and blood "others." You may hate the contents of the political sign on Mr. Jones' lawn, but you know he runs a pretty good hardware store and he sweeps his sidewalk every Sunday. Likewise, you may give a thumbs-up to the bumper sticker on Ms. Smith's car, but you know she's a careless driver who barely managed a "Sorry!" when she almost ran into you. Mr. Jones and Ms. Smith, in other words, are embodied human beings who arouse a complex set of reactions. But in [the] cyberspace, where we lack the body language, facial expressions, voice inflections and other cues that we use to get a read on someone, it's easy to reduce people to caricature. Politics becomes a way to attach a social identity to the disembodied creature of cyberspace. Mr. Trump Supporter or Ms. Progressive: that's all we know.

The above citation suggests that some critics are well disposed to anchor their belief in the dehumanisation effect of digitisation in the absence of human bodies or the absence of physicality of humanness on the cyberspace. Other dehumanisation-based critiques of digital diplomacy have rather explored the fact that the emergence of automation and computation and their subsequent application in the sphere of public diplomacy have made conditions favourable for diplomat-robots interactions and even the "robotisation" of diplomatic communications. To many critics, such a development strongly gives credence to the dehumanisation theory. In an opinion piece published in the online tabloid *Diplo*, digital diplomacy scholar Kurbalijah (2020) reviews ways in which Artificial Intelligence (AI) and computation technologies are making serious inroads into the practice of diplomacy causing the idea of automated diplomats to more and more become a popular culture or at least, a thinkable reality of the near future. In his review of ways in which automation is deployed in the conduct of public diplomacy, Kurbalija (2020) underscores China's constant leverage of the AI technology to provide insights for its diplomats into the possible scenarios and the evolution of events on the international arena. He also explores the use of AI in supporting economic diplomacy in trade negotiations.

The robot-diplomat interactions in particular is seen by many conservative and futurist critics as the worse form of dehumanisation of digital diplomacy following the belief that in any context of machine-human interaction, a robot is technically speaking, likely to treat its human counterpart as a machine or a similar automated system (Kurbalijah 2020; Shwab 2019). As Saldi (2018, p. 6) rightly puts it "for a machine, a human person, just like everything else, is only a set of numbers, is only one being among others, interchangeable, and an object of application of certain rules or protocols". What even makes the situation alarmingly dehumanising or "inhuman", according to Saldi is that "the delegation of powers to autonomous machines puts us [humans] on the path of negation, oblivion or contempt for the essential characteristics unique to the human persons".

It is hard to find empirical or unscientific studies that have captured African diplomats or African diplomacy researchers' positions specifically on the (perceived) dehumanisation effects of the digital technologies. However, the available literature reveals various cases of cultural or conservative critics who decry cybercultures' potential to dehumanise or pervert various aspects of life and work in Africa. In his book titled *Globalisation and cyberculture. An Afrocentric perspective,* Kehbuma (2016) notes for instance that the advent of digital technologies in Africa and their application in various sectors of life in the continent have had the awful effect of not only de-Africanising, but also removing humanness from the ways most modern African citizens communicate and interact among themselves. Kehbuma anchors his thesis in the popular belief that social media-assisted interpersonal communications do not integrate issues such as the sense of touch, smell, taste as well as the humanly generated non-verbal cues and the in-person characteristic which, according to him, are indexical of humanness and also, are typical of the African communication habits. He writes that:

When humans used to communicate through postal mail, telephones, and facsimile, the sense of touch, smell, taste, and feelings were sacrificed as a result of distance. They were only palpable by keeping hard record copies of them and generations to come had access to these records. Migration is as old as the earth itself, but never in human history has the sense of togetherness been so alien as we now have in the age of digital communication. Humans are now sharing digital spaces on the Internet and texting or blogging to one another and that has caused a strain in the in-person world as well as the virtual world as a result of multiple spaces and identities. (Kehnuma 2018, p. 97)

With close reference to digital diplomacy, a Kenyan diplomat (cited by Waithaka 2018) similarly decries the "un-human" character of digital diplomacy stressing the (un)realistic character of the concept of digitisation of diplomatic, particularly when conceptualising it in practical terms. This diplomat actually contends that "diplomacy cannot be skyped or teleconferenced. Diplomacy is shaking hands, looking straight in the eye, checking the body language and sharing a cup of coffee. In the same vein, we cannot expect diplomacy to be conducted in websites or social media platforms" (cited in Waithaka 2018, p. 113). At first sight, one may thing this diplomat has a misconception of digital diplomacy and humanness. However, a second look at things will definitely reveal that digital cultures in general and e-diplomacy in particular have, like other non-digital technology-based paradigms that emerged before them, challenged Africans to reinterpret or rethink their humanness.

In line with all the arguments reviewed above, one may plausibly say, digitalisation has the potential to dehumanise African digital diplomacy especially if it entails the application of such technological innovations as artificial intelligence, robotics, computation and other paradigms which fuse technologies and blur the lines between the physical, digital and biological spheres. However, it must be underlined that African diplomacy has not always been "humanised" from the physical and biological points of view. There have existed situations in which negotiations and intermediations in the African socio-political system have involved what could be regarded as paranormal diplomatic players. For instance, Africans tend to construe diplomacy as a normal part of life and death. This belief is seen in the fact that, in many parts of the continent, the traditional practice of "negotiating" with ancestors (spirits or non-humans) when need arises, still holds sway. In Swaziland for instance, people have the culture of practicing the *Incwala* which is a ritual involving some form of "negotiation" with the ancestors in order to ameliorate the welfare of the nation. Besides showing the seamless and non-linear nature of African traditional "diplomacy", this type of cultural practices illustrates the fact that Africans have since conceived or conceptualised diplomacy as a venture which may involve non-human players (spirits). Virtual diplomacy involving robots or non-human players should therefore not be viewed strictly as being too jarring since it somewhat has few things in common with what this author calls "spiritual African diplomacy", as exemplified by the *Incwala* practiced in Swaziland. Perhaps the only difference between the African practice of "spiritual diplomacy" and the modern robot-assisted diplomacy is that the latter involves machines while the former involves spirits or the spirit world. The latter is practiced on the cyberspace while the former happens in the spiritual space or the spiritual realm.

DIGITISATION AND THE MYTH OF THE DE-AFRICANISATION OF PUBLIC DIPLOMACY

A number of conservative critics are of the persuasion that, in both origin and essence, digital cultures are western or at least exogenous to Africa. Babalola (2019) has decried the fact that most Africans tend to construe digital cultures as a complex and ever-changing western phenomenon which Africans still need to fully appropriate. In the same line of thought, Kehbuma (2016) has contended that "as it stands, the cyberspace public sphere activities follow canons crafted and executed in the west by Westerners with little or no input from other non-developed regions of the world like Africa" (121). It is also not uncommon to find diplomats who share the above views and who, as a result, associate digital diplomacy with a western paradigm or a form of westernisation of African diplomatic practices. In her study of staffers at Kenyan MFAs, Waithaka (2018) notes the positions of various Kenyan diplomats who describe the digitalisation of diplomacy as a cultural revolution led by the west and an idea still not fully understood and/or accepted by African diplomats and intelligentsia.

Following the arguable belief that techno-cultures all have a western origin, many conservative Afro-centric critics have entrenched the tradition of describing digital cultures as yet another set of factors contributing to the proliferation of Euro-centric concepts and Americanism in Africa. A case in point is Kehbuma (2016) who argues that digital cultures are in their entirety agents of de-Africanisation on the continent as well as systems which subtly naturalise African dependency on the West for cultural models. He writes that:

Dependency is the tapeworm that keeps crawling through the veins of Africa to make her look up to Europe and America for communication help. It would appear that in-person communication that characterized communication between Africans prior to the arrival of the west was not effective so too were the drums, the bells, and the gongs to widen communication geo-cultural scope for all Africans. New media emergence from the west has become the welcome relief. That relief has infringed on African cultural space and swept her off her feet to the extent that the young and the old, including men and women, have fastened their seatbelts on the flight to adapt African culture in any given form on the new media communicative platform. (57)

It will not be surprising that, in line with the above essentialist view of cybercultures and African cultures, a number of critics associate digital diplomacy – which is an egregious example of the cyberculture – with Eurocentric concepts and agents of de-Africanisation on the continent. Such critics (as noted by Waithaka in her 2018 study of Kenyan diplomats) will likely view digital diplomacy as an agent of the proliferation of a variety of western paradigms in the practice of public diplomacy in Africa.

However, rather than hastily generalising, it will be more expedient to examine ways in which digitisation may facilitate the upholding of specific African cultural values in the sphere of public diplomacy. This paper focuses specifically on two such cultural values namely (i) predilection for collective and the prioritization of community rather than individuals and (ii) unhurried decisions. The first set of values do not run counter to the practice of digital diplomacy while the second rather appears incompatible to most of the savvy and pragmatic practices in digital diplomacy.

The Predilection for the Collective and the Prioritisation of Community Rather Than Individuals

As cultural values, these two concepts are observed among ethnicities and cultural groups across the African continent. In Southern Africa for instance, the two values find expression in such concepts as ubuntusim[1]. The upholding of these two values is very much compatible with best practices in digital diplomacy. In their leverage of social media for "Diaspora diplomacy" and "networked diplomacy", African MFAs and embassies can apply these values and ameliorate their conduct of public diplomacy. "Diaspora diplomacy" is a foreign policy strategy whose principal objective is to generate loyalty towards the home country and ultimately convert this loyalty into a political leverage.

It is conducted through a variety of instrumentalities some of which include the propagation of specific nationalistic and religious concepts and rhetoric and the organisation of cultural festivities and presidential elections abroad.

The social media can help African nations to maintain close and healthy relations with their global Diaspora. This can be done through the creation of social media-driven platforms that serve to inform their global diasporas about issues related to MFAs and embassies' activities abroad as well as information about domestic political events. Such platforms could also be used to provide their countries' diasporas with information about possibilities to invest at home. Thus, Diaspora diplomacy may enable African countries to support the formation of diaspora organisations abroad which contribute to retaining a collective identity.

Networked diplomacy may similarly enable the application of African concepts of solidarity and communalism. African diplomats and embassies may use the social media or their online activity to leverage their position in the global arena. Very active African MFAs and embassies may, for instance, deploy their tweets and Facebook messages to attract a significant number of their peers and for example, get them support a common cause. The sites of African MFAs and embassies may therefore become kind of information junctions that attract their peers or African counterparts and ultimately enable collective stance on issues involving or affecting the continent.

Unhurried Decisions

In the traditional African culture, leaders are expected to be wise and vested with practical and empirical knowledge. These wisdom and knowledge often manifest in the habit of being meticulous and careful when taking decisions that will affect the well-being of the nation. Unhurried decision-making mechanisms that involve consultations and consensus are definitely most valued (Awoniyi 2015; Wingo 2017). If unhurried decisions constitute a virtue and a practical tool deployed by African traditional leadership to manage delicate situations, in African MFAs or embassies, they are often the product of the top-down processes of governance. In the sphere of public diplomacy in particular, such top-down processes have mostly functioned as a factor disenabling prompt actions to deal with some uneasy situations. As such unhurried decisions have sometimes proved non-pragmatic particularly in context of uncertainties. In the present era of post truth, MFAs and embassies are compelled to react to events on the fly. They have to get ahead of events and act fast to avoid communication gaps that may lead to misunderstanding, negative perception of MFAs' activities and negative reactions from foreign publics. On this basis unhurried decisions may not always be the best solution.

CONCLUSION

Two key issues have been addressed in this paper. First, the paper has critically examined the myth stipulating that digitalisation contributes – or is likely to contribute – to the dehumanisation of African diplomacy. It has been argued in this paper that the progressive application in public diplomacy of such paradigms that fuse technologies and blur the lines between the physical, digital and the biological spheres actually gives some credence to the de-humanisation theory. In other words, by applying such technologies as automation, artificial intelligence, computation and robot-driven paradigms in the name of digitalisation, public diplomacy may be seriously dehumanised, especially if we conceive dehumanisation as removing or considerably reducing humanness or human touch from a human activity. The paper however also argued that de-humanisation if conceived along the definition provided above, should not really be seen as a novelty in African diplomacy sine African have in different period of their evolution conceptualised diplomacy as a practice that could involve para-normal and "unhuman" players. The act of negotiating with non-human (spiritual) entities such ancestors and gods is evidence that African diplomacy has not always been humanised. Also, the paper has argued that digital cultures including e-diplomacy have emerged to challenge humans to rethink or re-interpret their humanness.

The second issue addressed in the paper had to do with the purported de-Africanising effects of the digitalisation of African diplomacy. The paper argued that digitalisation is compatible with some African cultural values and incompatible with others. The use of social media driven communication to engage African diasporas and African counterparts can enable African MFAs uphold such values as predilection for the collective and the prioritisation of the community over the individual. Meanwhile, the African cultural value of unhurried decisions may rather be a threat to good practices in digital diplomacy. The dark side of the Internet has these last years compelled diplomats to act on the fly by getting ahead of events.

REFERENCES

Awoniyi, S. (2015). African cultural values: The past, present and future. *Journal of Sustainable Development in Africa, 17*(1), 1–13.

Babalola, T. A. (2019). Decolonising digital humanities: The African perspective. In L. Elisabeth & J. Wernimontr (Eds.), *Bodies of information: International feminism and digital humanists* (pp. 307–313). University of Minnesota Press.

Baker, T. (2016). Bringing the human back to work. *HR Daily Community*. https://community.hrdaily.com.au/profiles/blogs/bringing-the-human-being-back-to-work

Bjola, C. (2018). *Diplomacy in the digital age*. Real Institute/ Elcano.

Digital Diplomacy Blog. (2018). How diplomat can combat digital propaganda. *Exploring digital diplomacy.* https://digdipblog.com/2018/06/25/how-diplomats-can-combat-digital-propaganda/

Gluck, A. (2015). *De-Westernisation. A key concept paper*. University of Leeds.

Gordon, L. (2020). Humanism: Africa. *Encyclopaedia of African philosophy*.

Haslam, N. (2006). Dehumanization: An integrative review. *Personality and Psychology Review, 10*(3), 252–264. doi:10.120715327957pspr1003_4 PMID:16859440

Idang, G. E. (2015). African culture and values. *Phronomon, 16*(2), 97–111. doi:10.25159/2413-3086/3820

Kehbuma, L. (2016). *Globalisation and cyberculture. An Afrocentric perspective*. Springer.

Kurbalija, J. (2020). Negotiating with robot: Future or fiction? *Diplo*. https://www.diplomacy.edu/blog/negotiating-robots-future-or-fiction

Laverty, A. (2012). *The norms of African diplomatic culture: Implication for African integration*. London: IR 503

Linvingstone, S. D. (2011). *Less than human: Why we demean, enslave and exterminate other*. St. Martin's Press.

Louw, R. (2010). Africanisation: A rich environment for active learning on a global platform. *Progressio, 32*(1), 42–54.

Manor, I. (2018). *The Digitization of public diplomacy*. Palgrave.

Masolo, D. A. (2018). *Humanity: African thought. Encyclopedia of Science and Philosophy. Heterodyne to Hydrazoic Acid*. Science Rank.

Murithi, T. (2006). African approaches to building peace and social solidarity. *Accord: African Journal of Community Relations, 2*, 1–14.

Olusanya, G. (2012). 'Nigeria's role in the OAU'. *World event, 2*, pp.25-29.

Saldi, S. A. (2018). 'Editorial'. In De la Rochefocould A and Saldi S. (eds). The humanisation of robots and the robotisation of humans. Ethical reflections on lethal authonomous weapon systems and augmented soldiers. Chambesy: The Caritras in Veritate Foundation.

Shwab, K. (2019). *The fourth industrial revolution: What it means. How to respond*. World Economic Forum. https://www.weforum.org/agenda/2016/01/the-fourth-industrial-revolution-what-it-means-and-how-to-respond/

Spies, K. Y. (2018). African diplomacy. In G. Martel (Ed.), *Encyclopaedia of diplomacy* (pp. 1–14). John Wiley and Sons. doi:10.1002/9781118885154.dipl0005

Unver, A. (2017). *Computational diplomacy*. EDAM.

Waithaka, I. N. (2018). *Digital diplomacy: The integration of information and communication technologies in Kenya's Ministry of foreign Affairs 1964-2014*. [MA Thesis, University of Kenyatta University].

Wingo, A. (2017). Akan philosophy of the person. *Stanford Encyclopaedia of Philosophy* (1-14), Stanford: Stanford University Press.

Zaharma, R. S. (2018). *Battles to bridges: US strategic communication and public diplomacy after 9/11*. Springer.

ENDNOTES

1. Ubuntusim could be defined as humanism from an African perspective. The term actually refers to bonding and is in line with Zulu maxims which say that *I am because we are and I am human because I belong*. Ubuntuism is a manifestation of various tenets of the African philosophy which stipulate that an individual is human if he or she is member of the wider community.

Chapter 9
Social Health Protection During the COVID– Pandemic Using IoT

M. Aaschita Reddy
Sreenidhi Institute of Science and Technology, India

B Manidweep Reddy
Sreenidhi Institute of Science and Technology, India

C. S. Mukund
Sreenidhi Institute of Science and Technology, India

Kiran Venneti
Aditya College of Engineering, India

D. M. D. Preethi
iD https://orcid.org/0000-0002-6603-5252
PSNA College of Engineering and Technology (Autonomous), India

Sampath Boopathi
iD https://orcid.org/0000-0002-2065-6539
Muthayammal Engineering College, India

ABSTRACT

The agenda of this book chapter is to review existing technologies that aid societal health protection and recommend some possible approaches which will assist the mentioned scenario. Automations like big data and artificial intelligence (AI) deployed in healthcare sector can expedite pandemic response in ways that are strenuous to achieve all in all by humans. The sudden epiphany to trace COVID-19 in public has powered the innovation of data dashboards that visually unveil coronavirus epicentres. A cloud-based AI-assisted CT service is being engaged to differentiate pneumonia from the pandemic which dwindles risk factor in the present school of thought of the citizens worldwide. In conclusion, social health protection was an indispensable mechanism in prior to these challenging times and is escalating by prominence for delivering support to individuals during the crisis.

DOI: 10.4018/978-1-7998-8394-4.ch009

INTRODUCTION

The COVID-19 pandemic has engulfed the Indian subcontinent as well as affected people on a global scale, where it has caused damage not only physically but also socially and mentally. To tackle such problems, a new urge has surfaced to develop new innovations in this needed hour. In the wake of the pandemic, this book chapter is an ensemble effort in introspecting every aspect that is related to shielding the public's health through the deployment of various evolved and unacknowledged yet potential technologies that can aid the Earth in battling the present appalling circumstances.

Understanding social health in order to deploy efforts in fortifying the majority of the population against the virus's helix has become critical. In support of the above statement, for an individual's wellbeing, physical health is not the only determining parameter, one should also be mentally stable and active. The highlighted crisis has caused an unhealthy uplift in negative feelings, which has in turn affected mental health and greatly impacted the social ladder to the point where deaths due to mental stress and psychological breakdown are seen to compete with Coronavirus deaths. Taking the perilous reverberations of the pandemic into consideration, the call for an upsurge in spreading awareness of social health protection has become the main requisite, followed by elaborating on them in an outright manner, which will prevent such psychological complications.

The present epidemic has demonstrated that society can adapt and respond efficiently and adequately to any crisis owing to the advancements in science and technology. The brisk growth of cities in recent decades has turned them into pandemic epicenters, while their inhabitants have a higher risk of getting affected by coronavirus. Therefore, the eye of the world, at present is on developing advanced technologies to efficiently fight the crisis at hand. Although the COVID-19 pandemic is still ongoing, and it is too early to fully correlate the apparent edge of digital technologies to the pandemic's corresponding actions, it is an agreeable notion that the cornerstones of flattening the curve of COVID-19 cases are the paragon of technology and the conglomerated concept of social health. The emerging consensus is that smart healthcare plays an essential role in a comprehensive response to supplementing traditional social wellbeing strategies, and thus contributes to reducing the negative effects of COVID-19 on the human and economic aspects of a nation.

In this book chapter, we have simultaneously deliberated upon various facets of technology that have the potential to succor the world in these challenging times, as well as given the crux of the factors of society that are being overworked due to the ill effects of the virus in discussion.

BRIEF LAYOUT OF THE CORONA VIRUS ACROSS THE WORLD

In the moment, it has been just over a year since the start of the COVID-19 pandemic, which has claimed millions of lives, and hence, a question needs to be posed in this circumstance as to how the pandemic has ceased our normal lifestyle in the past 365 plus days. In this context, the pandemic has not only affected a few countries, but has affected every country in varying degrees. Over the last year, the global COVID-19 epidemic has been tracked using data on cases and deaths from around the world. We now know, however, that these only give us an incomplete view. The present epidemic emphasizes the need to priorities public health investments. The epidemic serves as a sobering reminder that early identification and prompt reaction by a robust and effective public health system may save lives and mitigate economic calamities. There is widespread expectation that the epidemic will act as a wake-up call, a catalyst for, or a forerunner of, positive change. For far too long, health has not been given the importance it deserves—public health has been chronically underfunded, with budget allocations hovering around 1.3 percent of GDP, resulting in inadequate basic health care. Health spending, particularly at the primary care level, must be significantly raised in the near future—much more than the 2.5 percent of GDP envisaged by the National Health Policy (2017) for 2022.

[1] The government of Bangladesh has broadened the national lockdown until May 30, 2021, but has loosened restrictions on long-distance public transportation, which had been prohibited since early last month. The wise decision was made after a significant decrease in the number of reported cases, deaths, and injuries. Data for verified COVID fatalities in 2020 is portrayed in the graph below.

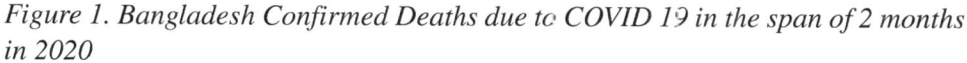

Figure 1. Bangladesh Confirmed Deaths due to COVID 19 in the span of 2 months in 2020

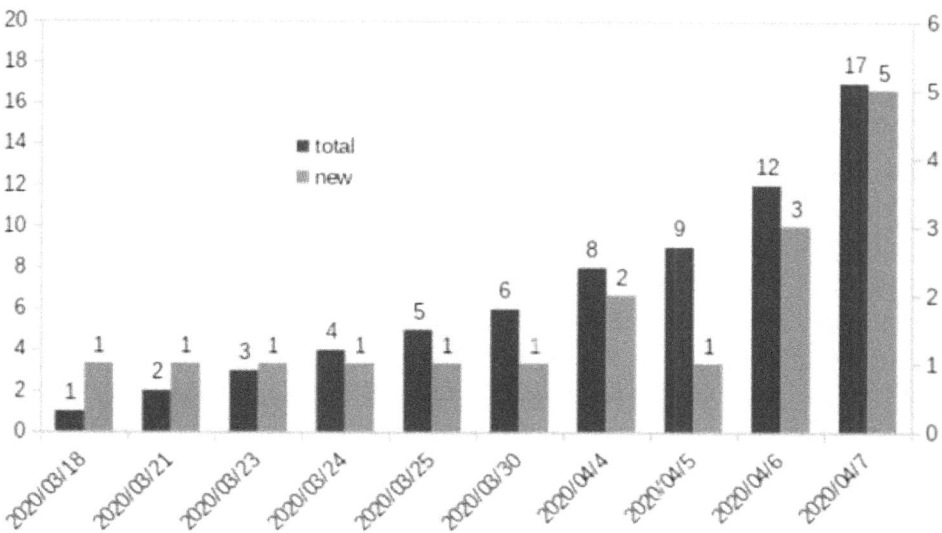

[2] One of the countries that has successfully responded to the pandemic's onset is the United States of America. As of Sunday, around 39 percent of the country's population had been immunized, and 49 percent had gotten at least one dose of the COVID-19 vaccine, according to the US Centers for Disease Control and Prevention. Vaccination rates, on the other hand, have been declining in recent times. An average of 1.8 million vaccination doses have been given out every day in the last seven days, down from a high of 3.1 million in April. Based on population, California, Oklahoma, and Nebraska had the lowest rates of infection.

Special Reference to the Outbreak of the Pandemic in Indian Subcontinent

In terms of surviving the pandemic's wrath, India is no different from other countries around the world. However, unlike other nations such as China, who had the virus under control far sooner, or the United States and the United Kingdom, who, though heavily impacted by COVID, had the financial wherewithal to preserve their people's livelihoods, India has suffered greatly.

The COVID-19 pandemic in India is part of a global coronavirus disease pandemic that commenced in 2019 and is caused by coronavirus 2 (SARS). On January 30, 2020, the first trial of COVID-19 in India, which originated in China, was reported. In Asia, India now has the highest number of confirmed cases. India had 26.7 million confirmed COVID-19 cases and 307,231 fatalities as of May 23, 2021, making it the world's second-largest number of confirmed cases (after the US) and third-largest number of COVID-19 deaths (after the United States and Brazil). Starting in April, it was projected that India's economy would swiftly recover and make up for the losses suffered in 2020. However, the country's large and uninterrupted increase in COVID cases, which has exposed the government's lack of preparation over the last year as woefully insufficient, suggests that India may not be able to make up for the economic (GDP) loss it suffered in the previous financial year (2020-21).

The response of India's government hospital system to COVID-19, in spite of the high death rate and other unpleasant circumstances, is a subject that receives little attention. The government has set aside Rs 15,000 crore as a crisis fund to relieve the aftermath of the COVID-19 treatment. As a supplement, after the unwise imports of protective kits from China were proven to be of poor quality, India decided to rely on domestic suppliers. With just 19,398 ventilators on hand, the government supported local medical equipment manufacturers by placing an order for 60,000 ventilators from local suppliers. However, this is a somewhat limited reaction. COVID-19's emergency response has shifted resources away from other vital health services, such as pregnancy and child health, with long-term effects.

Mrs. Nirmala Sitharaman, India's Finance Minister, also suggested increasing healthcare spending to Rs 2,200 crore to help improve public health systems and support a huge vaccination campaign to immunize 1.3 billion people. In total, the government projected Rs 5,540 crore for capital investment in 2021/2022, up 35 percent from the previous year's estimate. India spends about 1 percent of its GDP on health, one of the lowest percentages of any significant economy. The budget, according to Prime Minister Narendra Modi, is targeted at promoting "wealth and wellness" in a country that, after the United States, has the world's second biggest coronavirus caseload. A statistical analysis of the Indian Union Budget partitioned into various sectors is given below.

Figure 2. Major budget allocations (in Rs. crore) for various sectors of the Indian subcontinent

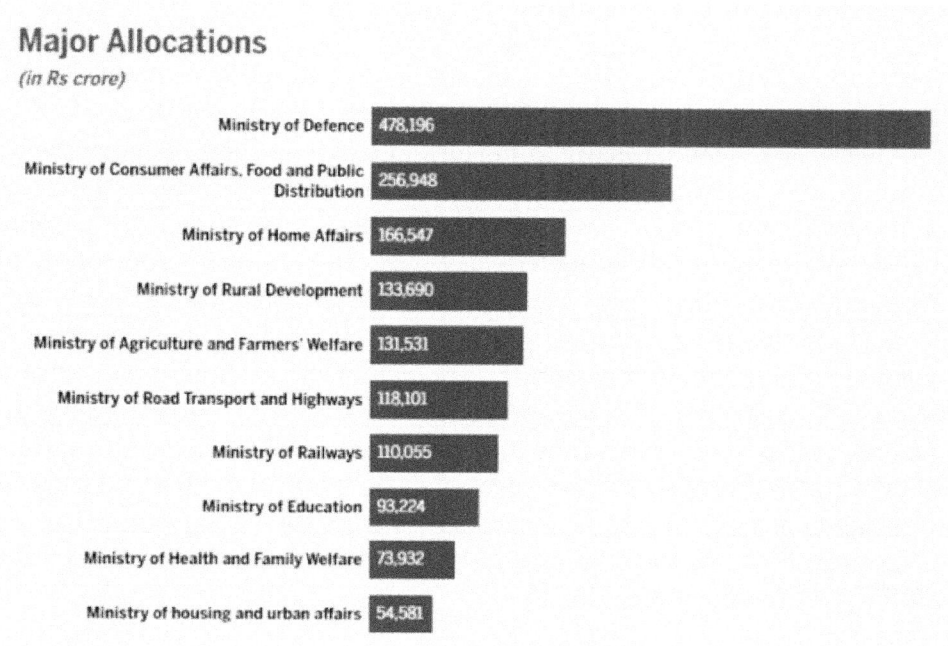

Despite the fact that the Indian government and administration are attempting to meet both ends in the context of the COVID-19 surge, [4], a number of small and medium Indian technology businesses have banded together to manufacture oxygen concentrators based on an open source design project called Marut, founded by robotic and automation experts. According to the project's partners, an oxygen concentrator with a flow rate of 10 liters per minute, a purity of over 93 percent, and the ability to operate around the clock could be supplied for half the price of imported devices, which cost between Rs 65,000 and 70,000 each. According to Technido, the pricing may be further reduced to Rs 35,000–40,000 per unit if 100 units or more are purchased in bulk.

EFFECTS ON THE ECONOMY DUE TO THE CATASTROPHIC HEALTH EXPENDITURES DURING THE PANDEMIC

With the COVID-19 pandemic causing havoc everywhere, India's public health system had many shortcomings, which had shown themselves by getting crushed by the overwhelming task of handling such a huge outbreak. However, expecting an already overburdened sector to perform this humongous task without government assistance is a foolish expectation. Furthermore, it is deeply regrettable that the country's private administrations of health institutions across the country charge exorbitant fees for admitting or providing access to basic medical amenities to COVID-19 patients. While continuing our review in this book chapter, the anguish of the middle class is also to be taken into account. Furthermore, we have elaborated on the paradigm of the present-day standpoint of the Indian subcontinent's social health system. We have also emphasized how respective governments must assist forces in strengthening pandemic response across their nation, as well as the role of the well-to-do sector in assisting the poor population in obtaining basic COVID-19 treatment amenities.

Figure 3. A pictorial abstraction of how the coronavirus has affected the economy

Agony of the Lower Middle Class and Poor Citizens in Relation to Medical Health Expenditure

It is more crucial than ever to have a healthy and inclusive economy that "leaves no one behind." Economic inclusion, on the other hand, is one of the most difficult issues planners face: altering the economic lives of the world's poor and severely poor to help them "graduate" from poverty. Over 700 million people worldwide are trapped in extreme poverty, especially in the face of climate change, warfare, or shocks like the COVID-19 epidemic. Poverty may be passed down through generations, leaving families with no prospect of escaping. One of the major life cycle concerns tackled by social protection is poor health: The financial burden of costly therapy is exacerbated by the ill worker's loss of income, and patients with inadequate cash may choose partial or no therapy, putting their health and lives in jeopardy.

While the middle class is frequently regarded as the glue that prevents modern free democratic economies from collapsing under the weight of ever-increasing inequality, with such a large, expected percentage of the impoverished middle class, the healthcare sector's inability to reduce an average person's bill remains unchanged. The present standards of the Indian subcontinent's social health paradigm are woefully poor, falling short of WHO standards. Based on the statistics, one out of every four Indians who are suffering from comorbidity with one of the diseases, such as coronavirus, is in jeopardy of dying from noncommunicable diseases before they reach the Indian average life expectancy of a human being, i.e., 60 years.

The new coronavirus has been such a disaster for the global economy for two key reasons: one, it infects everyone, regardless of their economic level, and two, it spreads mostly through human connection, just like economic development does. To put it another way, restricting the transmission of the virus and reducing the health-related harm entails incurring financial losses by definition. While this has been known since the beginning, it explains why so many governments and people have been caught off guard by the more destructive second and third waves. COVID-19 testing and treatment were mostly centered at government institutions during the early stages of the pandemic, and the ability of the impoverished to pay for testing at these institutions is completely ruled out.

In light of the plight of the lower income class of the population, the Indian administration must provide them with basic COVID care medical amenities at a low cost. It should be a conscious reality that the various administrations associated with the treatment of coronavirus must invariably curtail the amount that the infected patients need to spend in order to save their lives. In addition to the foregoing, governments must devise a contingency plan to ensure the availability of services to the general public, with a particular focus on those living below the poverty line, at an affordable cost. The upcoming topic of this book chapter sheds some light on how, despite the agony of the undernourished population across the world, in order to aid them, some well-known and high-status individuals have been stepping forward.

Aids Provided by the Upper Middle Class in Regards to the Pandemic

To create an equilibrium amongst the demarcated sectors of society as per their income, when the necessitous population extends a hand to accept aid in any form, the financially copious sector of society must supplement the former. While the cream of India comprises business giants and celebrities from various industries, in the recent rough crisis, the aforementioned people have been helping the poor population in financial and non-financial walks of life. The researchers proposed that the economic reaction to COVID-19 should be tailored to the public health response. More widespread and wide transfers are needed, for example, during heavy lockdown times when economic activity is severely restricted. In this situation, direct food aid can be very effective.

Targeting becomes considerably more crucial after lockdowns are released. Based on earlier research in Indonesia and Peru, which tailored programs—those focused on giving big per-capita benefits to disadvantaged households—appear to yield considerably better social welfare than widely targeted programmers, such as a universal basic income. When adopting targeted programmers, however, governments must be wary about accidentally omitting qualified individuals. Following the pandemic, the elite citizens are seen to have made direct donations or supported a non-profit organization that provides medical services and aids in the fight against the epidemic. In contrast, a significant portion of the elite has been fleeing to assist the health sector in this grave crisis. There aren't many people who are willing to give examples.

On the brighter side, celebrities, both Indian and foreign, such as assisting in any way they can: spreading the news, sharing COVID information, organizing events, and, of course, raising money to assist those who cannot help themselves. There's also comedian Kapil Sharma, who's contributing Rs 50 lakh to the Prime Minister's Relief Fund, as well as south Indian actors

Allu Arjun, Pawan Kalyan, and Rajinikanth. It has only been a few months since an innovative wearable device company, released a new smart band with numerous sensors that track human body temperature and other parameters. The world has witnessed one of the many acts of kindness in these unpleasant times: the company's brand ambassador, Mr. Akshay Kumar, has donated 1,000 of these bands to Mumbai Police personnel, who are the vanguards of the war against the virus in discussion.

Figure 4. These bar graphs together demonstrate the amount of coverage vs. the amount of expenditure by the public and private sectors

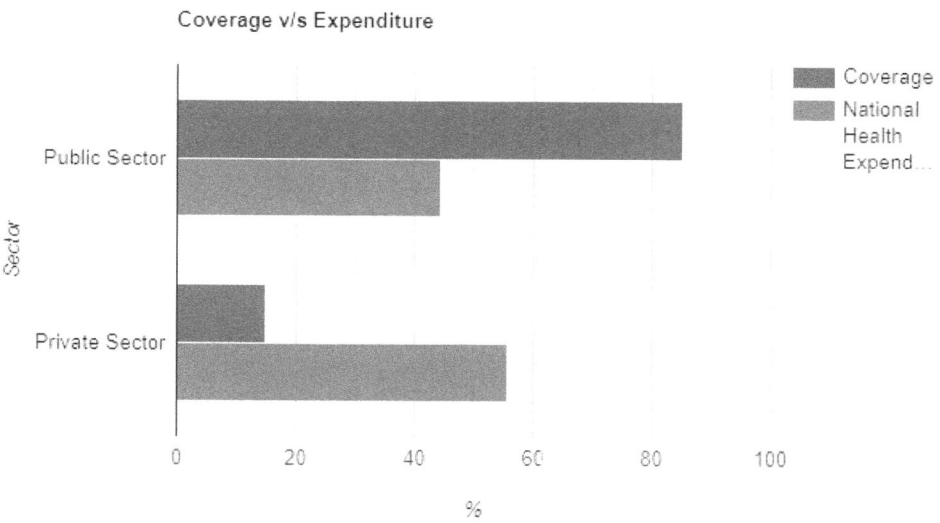

PARADIGM OF SOCIAL HEALTH DURING THE COVID-19

In the short term, the COVID-19 health crisis has morphed into a crisis of urban access, fairness, finance, safety, joblessness, public services, infrastructure, and transportation for many cities, all of which disproportionately affect the most disadvantaged in society. Deep-seated inequities, like where a person lives and works in a city, as well as gender and age, might cause the epidemic to have a disproportionate impact on groups that were previously vulnerable. The pandemic has already consumed an unacceptable number of loved ones for families across the nation. The consequences of this epidemic have been disastrous for the economic and social strata in society. It has forced millions of people into abject poverty. Many general survey participants expressed anxiety (44%) and worry (44.3%) in the context of the pandemic. Due to the overcrowding of public healthcare facilities, normal vaccines and follow-up care for people suffering from other diseases may have been interrupted (22.7 percent). The most common problem encountered as a result of the shutdown of places of entertainment and leisure was boredom (72.9 percent). There are two sides to every coin. When horrible things happen, good things happen, and the COVID-19 lockdown has resulted in constructive social developments. It has resulted in an improvement in society's human and social situations. With workplaces and academic institutions shutting down, it was necessary to move to a world of online learning and working from home. People became more conscious of the need for personal hygiene. We've all changed our lifestyles for the better, from covering our nose and mouth when we cough and sneeze to sanitizing our hands after handling anything else.

Fortifying the Vulnerable Sector of the Public Against Covid19 -Healthcare Staff, Elder Citizens, and Children

COVID-19 represents a 'moment of truth' for countries' social protection systems. Furthermore, the pandemic is the greatest challenge of our time, posing an existential threat to the health and livelihood of the world's entire population, yet affecting each age group and gender differently. In this topic, we have gathered the ways in which the coronavirus has impacted the most vulnerable population, providing a crux of the situation of frontline warriors, particularly those in the medical sector, senior citizens, and children.

Even though, thousands of health-care workers have been infected in Italy, the United States, and China, with many of them dying and the number of infected health-care workers is steadily increasing in the Indian subcontinent, the importance of local governments as front-line responders in disaster response, recovery, and rebuilding has been stressed during COVID-19. This is due to their leadership position in service delivery, infrastructure development, and urban resident mobilization. Under-resourced clinicians are facing unprecedented challenges as a result of the global pandemic. Healthcare staff, including doctors, nurses, medical cleaners, and many other health-care personnel, are among the sleep-deprived heroes. The courageous medical army, armed with thermometers, stethoscopes, and ventilators, remains firm in the war against coronavirus. While people in India and around the world are mainly confined to their homes, with businesses and educational institutions shut down in an attempt to contain the virus, doctors, health-care workers, and medical staff members are at the brim of the infection. They are our true heroes in these trying times, putting their own lives on the line with unselfish dedication for the sake of saving others. While they risk their own health, the health of their families, and, most crucially, their own lives, the least we can do is thank them for their efforts and cooperate by staying safe indoors. Infection control techniques, including universal precautions, must be taught to health care providers as soon as feasible, and an enabling environment must be provided, including an uninterrupted supply of personal protective equipment (PPE).

In and of itself, the COVID-19 disease has been more severe in the older population than in other age groups. Older citizens, especially those with a condition called comorbidity, are more likely to have underlying diseases like cardiovascular disease, diabetes, or respiratory sickness, all of which increase in prevalence. In Sweden, for example, 90 percent of COVID-19 deaths occurred in adults over the age of 70. In March, the Chinese Centers for Disease Control and Prevention released statistics revealing that the average COVID-19 case fatality rate for adults in their 60s was 3.6 percent, 8 percent for those in their 70s, and 14.8 percent for those aged 80 and up. In Italy, on the other hand, 40 percent of patients required hospitalization (Lazzerini & Ptoto, 2020), with roughly 7 percent requiring acute care. The fact that Italy's population is dominated by the elderly explains some of the country's high death rates. As a result, there is a strong message that the elderly and those with underlying comorbidities should be protected in society and at home, and that they should not be overexposed. Senior family members should avoid engaging with young people or children who appear to be healthy but are actually infected within the family.

In the face of this epidemic, public healthcare practices for children will be rudimentarily revised. Medical professionals, educators, and service providers are breaking free from their traditional ideologies and assisting every sector of the population that they can.

Innovations in systems that help at-risk children and families have been long overdue and are now more important than ever; such innovations will position us to provide higher value and better integrated care for all children in the future. Although children are not in the greatest danger of serious illness from coronavirus disease 2019 (COVID-19), required pandemic public health actions will have unforeseen implications for the nation's at-risk children's health and well-being. They observed that, because of COVID-19 fear, households report vaccination interruptions, and as a result, the overall vaccination rate for children turning one between March and May 2020 remains low. However, on the brighter side, there is a recent estimate stating that there will be a catch-up in the children who will take the vaccination against COVID-19.

Figure 5. Massachusetts COVID 19 cases by age group

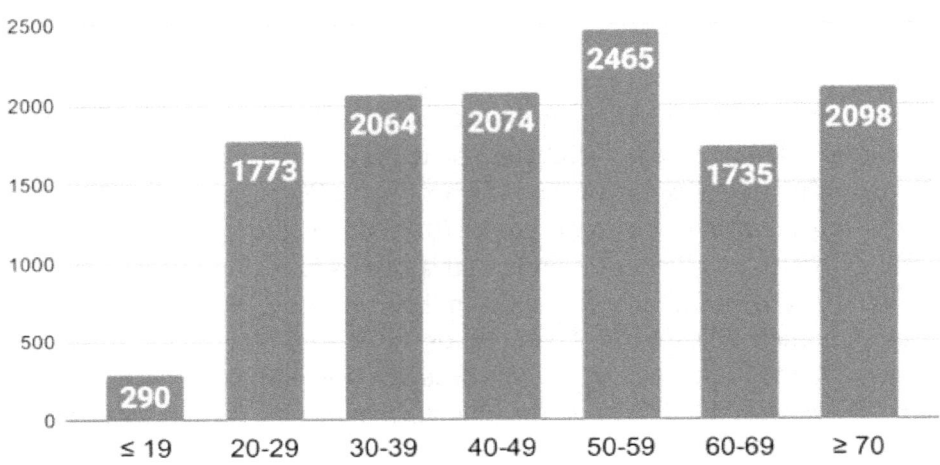

Data: Massachusetts Department of Public Health, as of 2020-04-05

Viewing the Impacts of the Pandemic Through a Psycho-Social Lens

Unwanted disasters like pandemics are known to have widespread psychological reverberations. The psychological impact may include the anxiety of the transmission of COVID-19, fatality fright, grief over the loss of loved ones due to the pandemic, isolation leading to distress that is completely unrelated to the biological changes caused by the infection of a virus, and mental effects in the population, amidst the existence of many life-consuming diseases. Contrarily, the social problems include pre-emergency social problems, including emergency political and financial subjugation; disruption of the order of societal health; and the hampering of togetherness. The nature of any threat is to primarily consume the vulnerable. It is a significant characteristic of COVID-19 that it impacts females, elderly citizens, children, and front-line warriors, unlike other sectors of the population. In this subtopic, we evaluate the socio-psychological strata that have been distressing recently.

The Deteriorating Mental Health of People Across the World During These Challenging Times

Pandemics have had a significant influence on the mental health of individuals who have been affected throughout history. Since millions of people have been forced to quickly adjust to new conditions and make drastic lifestyle adjustments, lockdown measures have heightened feelings of loneliness, isolation, restlessness, and anxiety among those who have not tested positive. While the majority of survivors and professionals exhibit a range of emotional reactions, only a tiny percentage of them suffer from diagnosable mental health issues. According to recent research, COVID-19 was found in 62,354 out of 69.8 million people in the United States, but psychiatric problems were found in nearly 20 percent of individuals identified with COVID-19 within three months after testing positive, including anxiety, sadness, and sleeplessness. One out of every four people polled had never received a mental health diagnosis. Studies on the psychosocial consequences of the SARS epidemic, for example, have revealed unfavorable psychological consequences such as higher depressive levels among people who were affected by the epidemic. Because of the huge and unexpected loss of life, the pandemic had a long-term impact on survivors' mental health, leaving many with a persistent sense of powerlessness and worry. The psychosocial burden of epidemics often exceeds the capability of people and communities to manage them, resulting in high levels of anguish that can endure long after the outbreak has stopped.

The Repercussions of Downfall of Women' Lifestyle During the Spread of Covid-19

Women in poverty are expected to be disproportionately affected by the pandemic, as they have been in the past with education, healthcare, and work. A new body of evidence indicates that females are less likely than men to die as a result of the pandemic due to specific circumstances. It's critical, whether these initial observations are true. Furthermore, we must ensure that COVID-19 has no effect on pregnancy outcomes and that SARS-CoV-2 is not transferred to the fetus or child through breast milk during pregnancy or labour. Into the bargain, we also need to shed light on the unintended consequences of the global lockdown on women's health in general and their psychological wellbeing. Furthermore, there are a number of issues concerning female sex in the context of COVID-19 and its associated circumstances that must be addressed. The tyranny of domestic violence against women, the alarming decline in women's mental health, the unacceptable denial to pregnant women of hospital visits and safe abortions have not received the attention they deserve during this pandemic period.

Before the epidemic, women were already 4 percent more likely than men to live in severe poverty. COVID-19 now threatens to widen the chasm even further. By 2021, the pandemic will have driven 47 million women and girls into poverty, raising the total number of women and girls living in poverty to 435 million. The epidemic also has the potential to undermine decades of progress in lowering the number of poor women. Women's poverty rates were expected to reduce by 2.7 percent between 2019 and 2021, but they are now expected to climb to 9.1 percent as a result of COVID-19. Even more so, 59 percent of disadvantaged women live in sub-Saharan Africa.

DEPLOYMENT OF SOCIAL HEALTHCARE TECHNOLOGIES TO PROTECT THE COMMON WELLBEING OF ALL SECTORS OF PUBLIC

Since the uprising of the COVID pandemic, it has wreaked havoc and caused drastic effects on a global scale. As a result, there was a push for the development of new technologies to combat the pandemic. However, the pandemic had not only caused damage physically but also mentally and socially, which affected the well-being of the people. In order to alleviate the ongoing public health problems, there has emerged a new sector for technological development— wearable devices for continuous monitoring of the user's important vitals. These concerns are addressed not only to the common people but to people of all sectors, economic standards alike.

Smart Healthcare to Revolutionize the COVID Hospitals

To attempt to comprehend how technology has helped to revolutionize the healthcare sector in the context of COVID 19, we must shed some light on the Internet of Things (IOT) and the Internet of Medical Things (IOMT), which have been by far extremely effective, along with the other technologies reviewed in this book chapter, in not only containing the spread of viruses but also easing the paradigm of patient care. The human era has moved a long way and seen many advancements in technology. In the advancement of technology, IOT and IoMT are two powerful weapons in the improvement of health care(Babu et al., 2022; Boopathi et al., 2022; Jeevanantham et al., 2022; S. et al., 2022; Samikannu et al., 2022; Sampath et al., 2022).

Internet of Medical Things: The application of the well-known Internet approach's fundamentals, ideas, tools, techniques, and concepts to the medical and healthcare sectors and domains is characterized as the "Internet of Medical Things" (IoMT).The proposed IoMT principles are crucial when medical services are required to be delivered in some remote locations. Healthcare, medical operations, and services have all been revolutionized as a result of the implementation of IoMT concepts and methodologies. Medical personnel, staff, healthcare professionals, and others must provide more impactful, productive, and successful services and treatments to their patients in light of the current COVID-19 pandemic. During the COVID-19 epidemic, this comprehensive study shows that employing an IoMT technique, it possible to give medical care to orthopedic patients. Orthopedic patients face various problems in this pandemic, including hospital visits and acquiring reports. The IoMT technique makes it easier and more effective to handle these challenges.

Internet of Things in Medical Care: The healthcare industry has been rapidly modernizing over the years, utilizing new technology to perform medical tasks with greater accuracy. Despite the COVID-19 outbreak, which severely interrupted people's daily lives around the world, technology has advanced at a much faster rate to aid the healthcare profession. A marketplace that helps businesses identify acceptable IoT service providers, is one of the many organizations that are assisting society during this pandemic. It has also expanded its services to aid the community's fight against COVID-19 by forming partnerships with a variety of other businesses that provide medical gadgets and equipment.

Use of Precautionary Technologies to Keep the Public Informed

Recognizing and describing the criteria allows administrators to select the technologies that best meet the needs of the organization. While acknowledging the significance of technology, it is vital to place a premium on the people who will use it. [9] A technology-first approach to smart hospital design is a pitfall that can lead to inefficient operations and, as a result, poor patient care.

Too frequently, the emphasis is on the technology rather than what it can do for the people who use it and are affected by it. As the coronavirus (Covid-19) outbreak spreads, new technological applications and efforts are emerging in an effort to keep the situation under control, treat patients effectively, and assist overworked healthcare workers while searching for new, effective vaccines. Based on the need of the hour, various technological domains have worked on new innovative applications. They also illuminate the primary legal and regulatory hurdles that hamper social and medical wellbeing.

Some of them are:

Artificial Intelligence: The way disease outbreaks are detected and handled has altered as a result of analytics, saving lives. The top brains across the borders of nations are putting their conglomerated efforts into developing contingency protocols to shield the world against the pandemic, which was initially discovered in Wuhan, China, in 2019-2020, but with a rhetorically digital approach. As the virus spreads, artificial intelligence is being widely used to follow the outbreak in real time, predict the future epicenters of the coronavirus, and pre-plan the precautionary ideologies for an efficient response. While the other real time applications of AI can make use of drone technology to transport medical aids, sterilize hospital rooms, it can also be used to find databases related to those drugs that could be effective against the virus. By using machine-learning algorithms and certain AI programmer, false information regarding COVID-19 can be determined. Social media giants like Facebook, TikTok, Twitter, and Google have teamed up with WHO to keep an eye on the false updates being passed on the internet regarding COVID-19.

Block Chain: Block chain mechanisms can be used to keep an eye on the spread of the virus by creating ledgers that are encrypted and updated to the latest number. Furthermore, when large amounts of real-time incoming data are published, such as during a virus outbreak, blockchain can reduce ambiguity, provide computational confidence, and serve as an automated platform for storing large amounts of data.

Figure 6. The above flowchart represents the different divisions of covid-19 technologies and their subsets

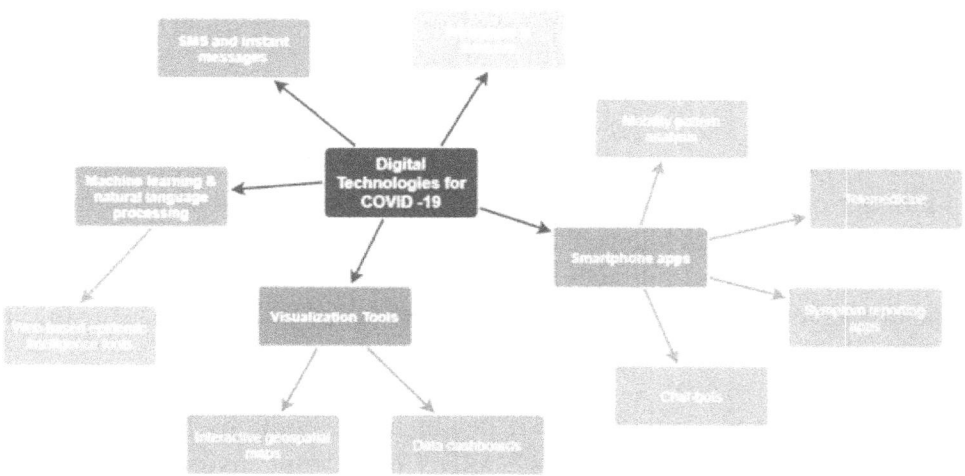

CHALLENGES ENCOUNTERED IN THE COURSE OF CONSISTENT USE OF THE DIGITAL TECHNOLOGIES IN THE EVENT OF A PANDEMICS

Limitations in the use of Internet of Things: Smart gadgets, wearables, and overall levels of connection and innovation in current medical equipment have irrevocably altered the business. And it's all for the best. However, in today's world, it's difficult to exaggerate the importance of the Internet of Things in healthcare. On the other hand, significant drawbacks associated with the widespread use of IoT in healthcare include:

1. The right to privacy may be jeopardized. Systems are compromised, as we've already said. Data security will demand a lot of attention, which will require substantial additional investment.
2. Centralized access is not permitted. There's a danger that dishonest intruders may get access to centralized systems and carry out their evil plans.

3.   Regulations governing healthcare across the world. International health organizations have already issued criteria that governmental medical institutions using IoT in their workflow must rigorously adhere to. These may restrict possible capacities to some extent.

4.   Drawbacks faced while the usage of Internet of Medical Things:

Although the emerging paradigm of the Internet of Medical Things has a number of vital applications for preventing epidemics and pandemics, it comes with a price. While constructing an IoMT system, interoperability to link devices from multiple sectors of medical units is a hurdle. Furthermore, doctors and healthcare professionals must be trained in order to use the IoMT system and applications efficiently.

5.   Breaches and flaws in social healthcare institutions that have been deliberately exposed by the COVID-19 pandemic include:

According to the nation's response to the ongoing epidemic, a state of chaos appears to be getting closer. The key barriers to India's COVID-19-containment plan have been insufficient infrastructure, a lack of human resources, insufficient financing, and large gender-based disparities in the access to medical amenities. According to the National Health Accounts, the Indian subcontinent's public health spending as a proportion of gross domestic product is just above 1 percent, making it one of the last counties in the race. Adding to the plight, millions of individuals are compelled to expect any affordable treatment from the unregulated private medical sector, which is already overburdened. According to the World Health Organization's 2017 health finance profile, India spends over two-thirds of its health budget on out-of-pocket expenses, whereas the global average is just 18.2 percent. According to government estimates, approximately 63 million Indians risk poverty each year owing to health-care expenditures alone.

In recent times, the need to comprehensively strengthen the medical infrastructure and the rudimentary preventive amenities against COVID-19 has become a matter of indispensable importance.

Figure 7. The above pie chart illustrates the amount of influence caused by the sources of the internet on people's healthcare decisions

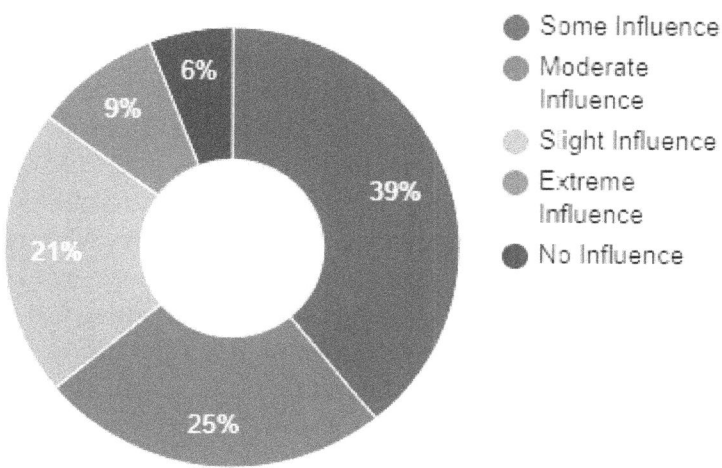

The influence of Internet resources on people's health-care decisions

- Some Influence
- Moderate Influence
- Slight Influence
- Extreme Influence
- No Influence

How the Dissimilarities in the Individual Situations of the Public Across the World Have Resulted in a Technological Divide Amongst Them

The World Health Assembly Resolution on Digital Health, adopted in 2018, acknowledged the importance of digital technology in achieving equilibrium in social healthcare systems. Despite the fact that trends are narrowing, there is still a digital divide, with more than half of the world's population not having access to mobile internet, making it nearly impossible for them to use any technology, particularly the social health protection technologies that are being discussed. The lack of access to smartphones is most noticeable in developing and poor countries, but it also affects people in high-income countries with lower socioeconomic status. According to the Pew Research Center, there are significant differences in mobile-communication access between persons aged 18–29 and those aged 50. There have also been reports of limited mobile internet connections, like in Myanmar, leaving certain communities ignorant of the epidemic and the efforts being made in the wake of the digital era in these trying times. Some populations, such as black and minority ethnic groups, have been disproportionately affected by this outbreak. As a result, it's critical to provide accessible tools and messages that may be adjusted to individual risks, languages, and cultural situations.

Our Take on Developing a Smart Wearable Device-Wearable COVID Companion

In the present COVID-19 crisis, quick wearable diagnostics are urgently deployed in order to contain COVID-19 infections as well as monitor and cease the infection control of coronaviruses. In the event of a pandemic, monitoring and tracking asymptomatic people is a matter of public health concern. In the event of a pandemic, we believe that such technology will be extremely useful in public health management in a country where manpower is considered a resource, like India. It will reduce the strain on testing facilities and health infrastructure by allowing asymptomatic individuals to be monitored remotely. Wearables continue to be the most popular item on the market. Apple and Android, two major mobile technology companies, are improving and upgrading their genuine wearables, providing additional health tracking functions. And the rest of the world isn't afraid to follow suit, resulting in an abundance of multi-purpose mini devices. we have attempted to pledge ourselves to lend our hands towards flattening the curve of COVID 19, and as a result,, we developed a wearable device analogous to a watch: Wearable COVID Companion. In this section of our book chapter, we imagined our Wearable COVID Companion, including its 3D model and block diagram, while explaining how it works.

Description

It is a smart wearable that captures information such as a person's temperature, pulse, and oxygen levels. These readings are then immediately transferred to a government-managed database through SMS or a server, where they may be seen by physicians or the public health agency. When a person's body temperature rises above acceptable levels or their oxygen saturation falls below acceptable levels, health professionals can be notified via SMS. As a result, it also acts as a precautionary device for others, as it scans for Bluetooth devices on a regular basis and stores their information, so that if any irregularities are discovered after updating the database, an SMS is sent to the person and others who have also been in contact (information from Bluetooth) to take precautions and isolate for quarantine.

Components Used

1. 2 SEEDUINO'S
2. RYB 080I(Bluetooth)
3. MAX030100(Oximeter sensor)
4. DS18B20(Temperature sensor)
5. SMS900(GSM MODULE)

Working Mechanism

This smart band contains the above components and functions as per the below block diagram:

Figure 8. Block diagram

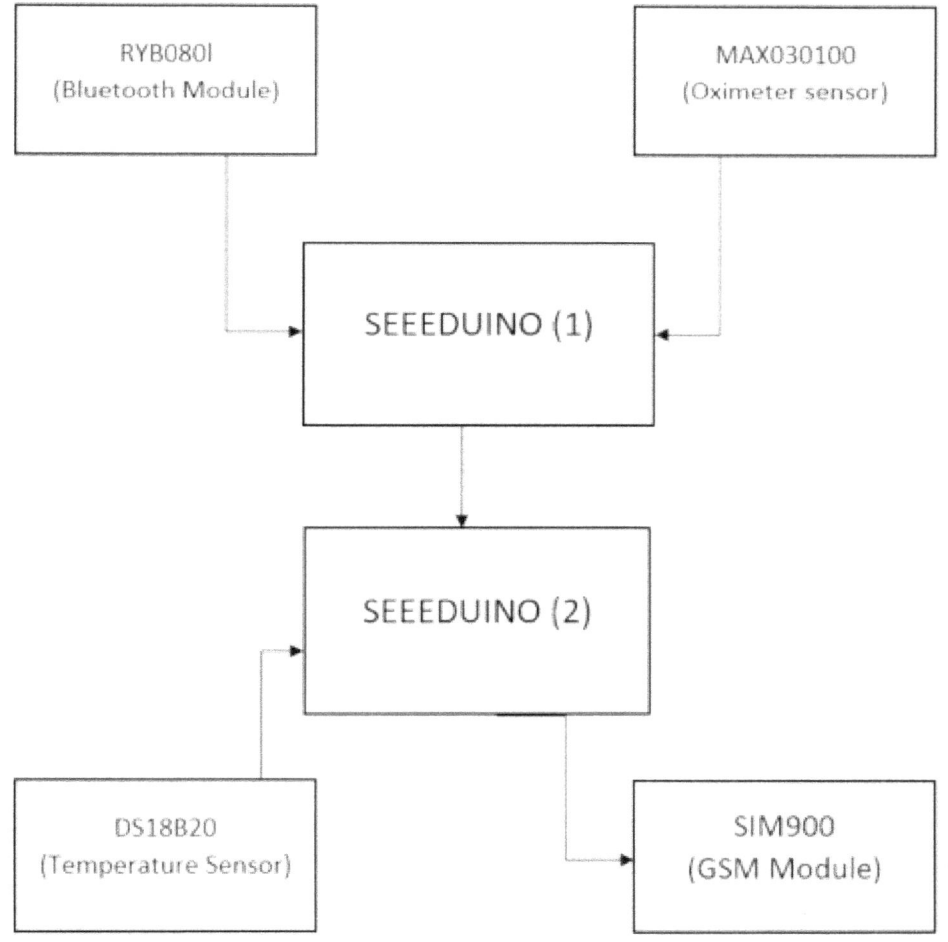

We used two SEEEDUINOs in this project, as shown in the image above, because the GSM and Bluetooth sensors each require a tx/rx protocol, and having only one SEEEDUINO would have caused interference with the tx/rx protocol. The SEEEDUINO (1) receives data from the oximeter (such as pulse and oxygen levels), while the Bluetooth sensor gathers data from persons who have come into contact with the individual wearing the smart band. All of this information is processed and transferred to SEEEDUINO (2), where temperature data is also captured and integrated with SEEEDUINO information (1). All of this information is delivered to the GSM module, which subsequently transmits it through SMS to a government-owned toll-free number, which is used to gather data and upload it into the database. The data in this database is separated by reference ID and personal information. If there are any inconsistencies in the data, the specific id will be flagged, and a protocol will be initiated in which various SMS will be sent to the person as well as the people who have come into contact with them to take precautions and isolate them (or) via server, which directly updates the database at the cost of a mobile data connection (internet).

We chose to upload data to the database through SMS as the default method since the government may give a toll-free number to which the data may be provided at no additional cost, and SMS will be received and notified whether or not a person has a smartphone.

3-Dimensional Design Models

This is an initial design and will be changed according to our ideation for different advanced versions of the Wearable COVID Companion.

ICT TOOLS

ICT scheme@ schools: In order to close the digital gap in India and give students in both rural and urban areas equal access, the information and communication technology scheme was created in 2004. To encourage ICT usage in rural areas, access devices, internet connectivity, and ICT literacy promotion are used. The administration of the Union Territories and the State Governments have teamed up to offer computer-assisted instruction in Secondary and Higher Secondary Government Schools.

Figure 9. Wearable COVID Companion

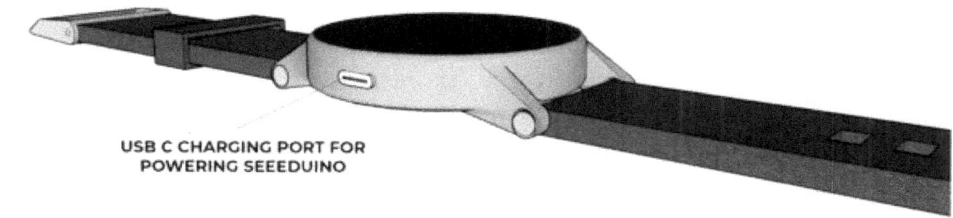

View of the Wearable COVID Companion's charging port.

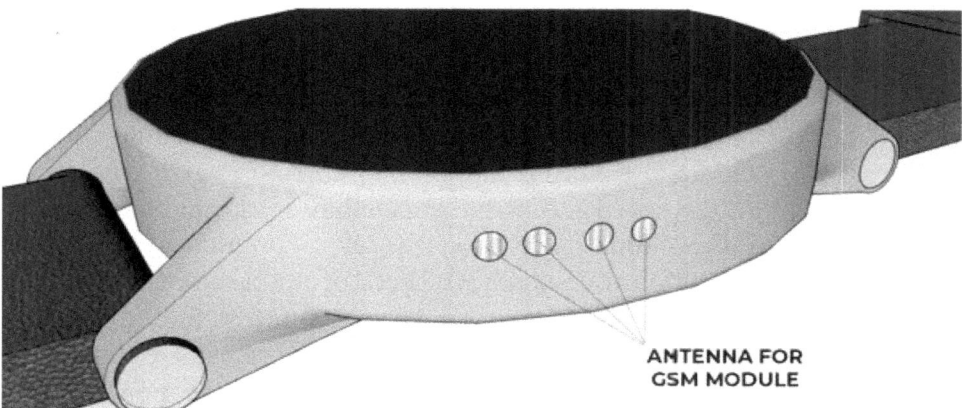

View of antenna for the GSM module for Wearable COVID Companion.

Top view of Wearable COVID Companion

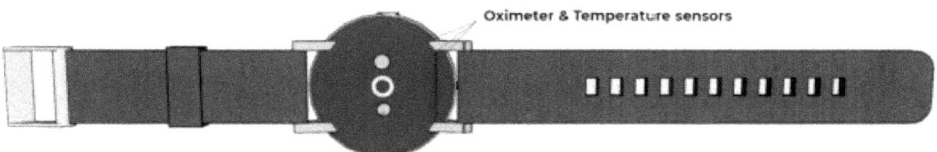

Oximeter and Temperature sensors in Wearable COVID Companion

SWAYAM (Study webs of Active-learning for Youth Aspiring Minds): The Ministry of Human Resource Development has launched a programme called MOOC to guarantee equality, accessibility, and high-quality education for everyone. For the greatest material, seven national coordinators have been chosen. IGNOU is for students who are not enrolled in school, and SWAYAM gives them the chance to broaden their horizons.

Online Labs: With the use of online labs, lab experiments can be taught more effectively and affordably, reaching students who might not otherwise have access to them. They create realistic lab conditions using cutting-edge simulation technologies and apply mathematical techniques to illustrate difficult operations. The Amrita CREATE, Amrita Vishwa Vidyapeetham, and CDAC, Mumbai-based educational programme known as The Online Labs provides experiments in the physical, chemical, and biological sciences for students in classes 9 through 12 with curriculum that is in line with the NCERT/CBSE and State Board Syllabi.

EDUSAT: ISRO launched the first Indian satellite for educational purposes.

Rationale of the Study: The modern day is dominated by digital technology, and education is a fundamental need of any civilised civilization. The development of technology literacy and new abilities, including problem-solving, teamwork, critical reading, and information retrieval, are considered as naturally occurring in the educational sector. ICT facilitates communication between theory and policy, practise and research, and pedagogy and technology.

ICTs have the power to improve learning environments, increase motivation to learn, and increase access to education, making them a comprehensive system of support for learning. ICT can be used to spread knowledge about and encourage the adaptation, adoption, translation, and distribution of scarce educational resources. It can also be used to digitise and distribute already-existing print resources, address teacher capacity building, improve efficiency, and create applications, software tools, media, and interactive devices to encourage students' and teachers' creative, aesthetic, analytical, and problem-solving abilities, and sensibilities.

RECOMMENDATIONS

With the appearance of a plethora of new technologies and emerging devices to combat COVID-19, the public gains hope for a more secure and peaceful future. On the contrary, no matter how strong and dependable the plans imposed by governments are, there will always be privacy violations and other minor public issues when they are implemented. Because leveraging the digitization era for social protection programmers can indeed be a potential fix, However, this swift flip to digital systems that are flawed in design and implementation might leave us with an inefficient system, which will just add to the pile of grievances. Hence, we still identify certain recommendations to be highlighted from our end. 1) Integration of Google Maps with the COVID database of areas affected will also be a useful precautionary technology, provided the databases are updated regularly and are accurate. This can be achieved by setting up a link between a person's Google Maps application and a relative database carrying the list of COVID hotspots. This can help an individual escape the epicenters via an alert mechanism and prevent any unnecessary infection spread. The IoT concepts are also applied in the agricultural field also. The hydroponics, aeroponics, and aquaponics forming methods were also applied using IoT concepts to improve the vegetable growth rate(Boopathi et al., 2023; Domakonda et al., 2022; Kumara et al., 2023; S. et al., 2022; Sampath et al., 2022; Vanitha et al., 2023; Vennila et al., 2022).

2)   Use of radio broadband networks for live updates of COVID affected regions:

Making use of mass media and FM radio can instantly protect people from entering COVID-affected areas. This can be done by radio stations whose broadband receivers (radios) can be set up in every public gathering place like railway stations, food courts, and vehicles (inbuilt radios).

CONCLUSION

On humanitarian grounds, COVID has changed the perspective of every single human being. It has urged all sectors to redefine their perception of social health and its immense importance. The virus that originated in 2019 has made us realize the importance of the social bonds we maintain, and put each one of us through a testing period. As a precautionary step towards the upcoming waves of the pandemic, it is the prime duty of every single individual to be aware of the precautions in order to prevent further damage in the current situation. The COVID-19 epidemic put enormous strain on health and human services (HHS) professionals and the government to subsidize the costs of nursing active and suspected COVID patients. However, it sparked the fast invention and acceptance of digital alternatives, which benefited both service consumers and practitioners but, in the majority, the general public.

This pandemic left a major stamp on all human beings and taught them about social awareness. The definition of the "new normal "way of living life will continue to exist if there is no improvement in the thought process of a human being with respect to his psychological thoughts, with special emphasis laid on how to shield themselves as well as the people in their immediate surroundings from the ongoing COVID-19. The aforementioned statement will exercise its way into ameliorating every nation's stance on social health protection and the associated technologies in the present crisis. Apart from the ill effects, COVID-19 has allowed the present generation to push beyond limits and also triggered them to scoop out new solutions and measures to combat such a pandemic, which was never seen in the entire history of social livelihood. Through this paper, we intend to give out some practically applicable solutions that may have the potential to stop the unwanted expansion of the pandemic.

It is a question fabricated after witnessing the recent moment of truth that might attract some light in the appalling situation: Is it possible to preserve the inventive sprit exhibited in 2020 beyond the crisis and pave the way for long-term change? Nonetheless, unlike earthquakes or tropical storms, the COVID-19 pandemic crisis will not be over in a matter of hours or days. It will remain with us for at least another year, and maybe several more. As a positive consequence, in addition to resolving urgent difficulties, there is also the potential to enhance the socio-economic strata and social healthcare sector of India and other nations across the globe.

REFERENCES

Al-Herz, W., Bousfiha, A., Casanova, J. L., Chatila, T., Conley, M. E., Cunningham-Rundles, C., Etzioni, A., Franco, J. L., Gaspar, H. B., Holland, S. M., & Klein, C. (2014, April 22). Primary immunodeficiency diseases: An update on the classification from the international union of immunological societies expert committee for primary immunodeficiency. *Frontiers in Immunology*, *5*, 162. PMID:24795713

Allam, Z. & Jones, D. S. (2020). On the coronavirus (COVID-19) outbreak and the smart city network: universal data sharing standards coupled with artificial intelligence (AI) to benefit urban health monitoring and management. In *Healthcare, 8*(1), p. 46. Multidisciplinary Digital Publishing Institute.

Babu, B. S., Kamalakannan, J., Meenatchi, N., Karthik, S., & Boopathi, S. (2022). Economic impacts and reliability evaluation of battery by adopting Electric Vehicle. *IEEE Explore*, 1–6.

Binks, A. P., LeClair, R. J., Willey, J. M., Brenner, J. M., Pickering, J. D., Moore, J. S., Huggett, K. N., Everling, K. M., Arnott, J. A., Croniger, C. M., Zehle, C. H., Kranea, N. K., & Schwartzstein, R. M. (2021, February 18). Changing Medical Education, Overnight: The Curricular Response to COVID-19 of Nine Medical Schools. *Teaching and Learning in Medicine, 33*(3), 1–9. doi:10.1080/10401334.2021.1891543 PMID:33706632

Boopathi, S., Arigela, S. H., Raman, R., Indhumathi, C., Kavitha, V., & Bhatt, B. C. (2022). Prominent Rule Control-based Internet of Things: Poultry Farm Management System. *IEEE Explore*, 1–6.

Boopathi, S., Siva Kumar, P. K., & Meena, R. S. J., S. I., P., S. K., & Sudhakar, M. (2023). Sustainable Developments of Modern Soil-Less Agro-Cultivation Systems. In Human Agro-Energy Optimization for Business and Industry (pp. 69–87). IGI Global. https://doi.org/ doi:10.4018/978-1-6684-4118-3.ch004

Brem, A., Viardot, E., & Nylund, P. A. (2021, February 1). Implications of the coronavirus (COVID-19) outbreak for innovation: Which technologies will improve our lives? *Technological Forecasting and Social Change, 163*, 120451. doi:10.1016/j.techfore.2020.120451 PMID:33191956

Budd, J., Miller, B. S., Manning, E. M., Lampos, V., Zhuang, M., Edelstein, M., Rees, G., Emery, V. C., Stevens, M. M., Keegan, N., Short, M. J., Pillay, D., Manley, E., Cox, I. J., Heymann, D., Johnson, A. M., & McKendry, R. A. (2020). Digital technologies in the public-health response to COVID-19. *Nature Medicine, 26*(8), 1183–1192. doi:10.103841591-020-1011-4 PMID:32770165

Chauhan, A., Jakhar, S. K., & Chauhan, C. (2021, January 10). The interplay of circular economy with industry 4.0 enabled smart city drivers of healthcare waste disposal. *Journal of Cleaner Production, 279*, 123854. doi:10.1016/j.jclepro.2020.123854 PMID:32863607

Debata, B., Patnaik, P., & Mishra, A. (2020, November). COVID-19 pandemic! It's impact on people, economy, and environment. *Journal of Public Affairs*, *20*(4), e2372. doi:10.1002/pa.2372

Dilshan, N. W. T. (2020). *A Workforce Perception Of The Impact Of Covid-19 (Novel Coronavirus) On Job Security Of Tourism Industry In Sri Lanka.* UOC Tourism.

Domakonda, V. K., Farooq, S., Chinthamreddy, S., Puviarasi, R., Sudhakar, M., & Boopathi, S. (2022). Sustainable Developments of Hybrid Floating Solar Power Plants: Photovoltaic System. In Human Agro-Energy Optimization for Business and Industry (pp. 148–167). IGI Global.

Galán, M., Velasco, M., Casas, M., & Goyanes, M. (2020). *SARS-CoV-2 Seroprevalence Among All Workers in a Teaching Hospital in Spain: Unmasking The Risk.* Cold Spring Harbor Laboratory.

Haleem, A., & Javaid, M. (2020, December 1). Medical 4.0 and its role in healthcare during COVID-19 pandemic: A review. *Journal of Industrial Integration and Management.*, *5*(4), 531–545. doi:10.1142/S2424862220300045

Javaid, M., Haleem, A., Singh, R. P., Haq, M. I., Raina, A., & Suman, R. (2020, December 1). Industry 5.0: Potential applications in COVID-19. *Journal of Industrial Integration and Management.*, *5*(4), 507–530. doi:10.1142/S2424862220500220

Javaid, M., Haleem, A., Vaishya, R., Bahl, S., Suman, R., & Vaish, A. (2020, July 1). Industry 4.0 technologies and their applications in fighting COVID-19 pandemic. *Diabetes & Metabolic Syndrome*, *14*(4), 419–422. doi:10.1016/j.dsx.2020.04.032 PMID:32344370

Jeevanantham, Y. A., Saravanan, A., Vanitha, V., Boopathi, S., & Kumar, D. P. (2022). Implementation of Internet-of Things (IoT) in Soil Irrigation System. *IEEE Explore*, 1–5.

Joshi, A. (2021, January 6). COVID-19 pandemic in India: Through psycho-social lens. *Journal of Social and Economic Development*, *23*(S2), 1–24. doi:10.100740847-020-00136-8 PMID:34720479

Joshi, A. M., Shukla, U. P., & Mohanty, S. P. (2008). *Smart Healthcare for Diabetes: A COVID-19 Perspective.* arXiv preprint arXiv:2008.11153.

Kaiser, M. S., Al Mamun, S., Mahmud, M., & Tania, M. H. (2021). *Healthcare robots to combat COVID-19. InCOVID-19: Prediction, Decision-Making, and Its Impacts.* Springer.

Ko, C.-H., Yen, C.-F., Yen, J.-Y., & Yang, M.-J. (2006). Psychosocial impact among the public of the severe acute respiratory syndrome epidemic in Taiwan. *Psychiatry and Clinical Neurosciences, 60*(4), 397–403. doi:10.1111/j.1440-1819.2006.01522.x PMID:16884438

Kumara, V., Mohanaprakash, T. A., Fairooz, S., Jamal, K., Babu, T., & B., S. (2023). Experimental Study on a Reliable Smart Hydroponics System. In *Human Agro-Energy Optimization for Business and Industry* (pp. 27–45). IGI Global. https://doi.org/ doi:10.4018/978-1-6684-4118-3.ch002

Mujawar MA, Gohel H, Bhardwaj SK, Srinivasan S, Hickman N, & Kaushik A. (2020). Aspects of nano-enabling biosensing systems for intelligent healthcare; towards COVID-19 management. *Materials Today Chemistry.*

O'Reilly-Shah, V. N., Gentry, K. R., Van Cleve, W., Kendale, S. M., Jabaley, C. S., & Long, D. R. (2020, May 12). The COVID-19 pandemic highlights shortcomings in US health care informatics infrastructure: A call to action. *Anesthesia and Analgesia, 131*(2), 340–344. doi:10.1213/ANE.0000000000004945 PMID:32366769

Osetskyi, V. (2019). IoT in healthcare: Use cases, trends, advantages and disadvantages. *Existek.* https://existek.com/blog/iot-in-healthcare/

Raj, S., Abu-Ghname, A., Davis, M. J., & Maricevich, R. S. (2020, September 1). The COVID-19 Pandemic: Implications for Medical Students and Plastic Surgery Residency Applicants. *Plastic and Reconstructive Surgery, 146*(3), 396e–397e. doi:10.1097/PRS.0000000000007232 PMID:32541533

Rogers, R. (2021). Internet of Things-based Smart Healthcare Systems, Wireless Connected Devices, and Body Sensor Networks in COVID-19 Remote Patient Monitoring. *American Journal of Medical Research (New York, N.Y.), 8*(1), 71–80. doi:10.22381/ajmr8120217

Russell, W., Austin, G. L., Barton, K., Nugent, N., Kerr, D. S., Neil, R. S., & Lee-Lawrence, T. A. (2021). *'Quality'Response to COVID-19: The Team Experience of the Office of Quality Assurance, University of Technology.* CCEAM.

S., P. K., Sampath, B., R., S. K., Babu, B. H., & N., A. (2022). Hydroponics, Aeroponics, and Aquaponics Technologies in Modern Agricultural Cultivation. In *Trends, Paradigms, and Advances in Mechatronics Engineering* (pp. 223–241). IGI Global. https://doi.org/ doi:10.4018/978-1-6684-5887-7.ch012

Sabbati G, Tkalec I. Living in the EU: Work before the coronavirus crisis.

Sachs, J. D., Karim, S. A., Aknin, L., & Allen, J. (2020). Lancet COVID-19 Commission Statement on the occasion of the 75th session of the UN General Assembly. *The Lancet.*

Samikannu, R., Koshariya, A. K., Poornima, E., Ramesh, S., Kumar, A., & Boopathi, S. (2022). Sustainable Development in Modern Aquaponics Cultivation Systems Using IoT Technologies. In *Human Agro-Energy Optimization for Business and Industry* (pp. 105–127). IGI Global.

Sampath, B. C. S., & Myilsamy, S. (2022). Application of TOPSIS Optimization Technique in the Micro-Machining Process. In Trends, Paradigms, and Advances in Mechatronics Engineering (pp. 162–187). IGI Global. doi:10.4018/978-1-6684-5887-7.ch009

Sherry, S. T., Ward, M. H., Kholodov, M., Baker, J., Phan, L., Smigielski, E. M., & Sirotkin, K. (2001, January 1). dbSNP: The NCBI database of genetic variation. *Nucleic Acids Research, 29*(1), 308–311. doi:10.1093/nar/29.1.308 PMID:11125122

Tavakoli, M., Carriere, J., & Torabi, A. (2020, July). Robotics, smart wearable technologies, and autonomous intelligent systems for healthcare during the COVID-19 pandemic: An analysis of the state of the art and future vision. *Advanced Intelligent Systems., 2*(7), 2000071. doi:10.1002/aisy.202000071

Vanitha, S. K. R., & Boopathi, S. (2023). Artificial Intelligence Techniques in Water Purification and Utilization. In *Human Agro-Energy Optimization for Business and Industry* (pp. 202–218). IGI Global. doi:10.4018/978-1-6684-4118-3.ch010

Vargo, D., Zhu, L., Benwell, B., & Yan, Z. (2020). Digital technology use during -19 pandemic: A rapid review. *Human Behavior and Emerging Technologies.*

Vennila, T., Karuna, M. S., Srivastava, B. K., Venugopal, J., Surakasi, R., & Sampath, B. (2022). New Strategies in Treatment and Enzymatic Processes: Ethanol Production From Sugarcane Bagasse. In Human Agro-Energy Optimization for Business and Industry (pp. 219–240). IGI Global.

Virani, S. S., Maddox, T. M., Chan, P. S., Tang, F., Akeroyd, J. M., Risch, S. A., Oetgen, W. J., Deswal, A., Bozkurt, B., Ballantyne, C. M., & Petersen, L. A. (2015, October 20). Provider type and quality of outpatient cardiovascular disease care: Insights from the NCDR PINNACLE Registry. *Journal of the American College of Cardiology*, *66*(16), 1803–1812. doi:10.1016/j.jacc.2015.08.017 PMID:26483105

Whitelaw, S., Mamas, A., Topol, E., Van Spall, H. (2020). Applications of digital technology in COVID-19 pandemic planning and response. *The Lancet Digital Health*.

Wong, C. A., Ming, D., Maslow, G., & Gifford, E. J. (2020). Mitigating the Impacts of the COVID-19 Pandemic Response on At-Risk Children. *Pediatrics*, *146*(1), e20200973. doi:10.1542/peds.2020-0973 PMID:32317311

Compilation of References

Abbas, J., Wang, D., Su, Z., & Ziapour, A. (2021). The Role of Social Media in the Advent of COVID-19 Pandemic: Crisis Management, Mental Health Challenges and Implications. *Risk Management and Healthcare Policy*, *14*, 1917–1932. doi:10.2147/RMHP.S284313 PMID:34012304

Abdurahmanlı, E. (2021). Definition of diplomacy and types of diplomacy used between states. *Anatolian Academy Social Sciences Journal*, *3*(3), 580–603.

Adam, G. (2020). China's failed pandemic response in Africa. *Foreign Policy Essay*. https://www.lawfareblog.com/chinas-failed-pandemic-response-africa

Adesina, O. (2022). *Africa and the future of digital diplomacy. Foresight Africa: Technological innovations: Creating and harnessing tools for improved livelihood*. Foresight Africa.

Adesina, O. S. (2017). Foreign policy in an era of digital diplomacy. *Cogent Social Sciences*, *3*(1), 1–13. doi:10.1080/23311886.2017.1297175

Adewale, M. P. (2014). Nigeria's China connection. *New York Times*. https://www.nytimes.com/2014/05/08/opinion/majapearce-nigerias-china-connection.html

Adler, E. (2019). *World ordering: A social theory of cognitive evolution*. Cambridge University Press. doi:10.1017/9781108325615

African Union. (2020, September 10). *The digital transformation strategy for Africa (2020-2030)*. AU. https://au.int/sites/ default/files/documents/38507-doc-DTS-english.pdf

Agency Report (2020, May 26). COVID-19: China places ban on foreigners to prevent second wave of infections. *Independent*. https://independent.ng/covid-19-china-places-ban-on-foreigners-to-prevent-second-wave-of-infections/

Agley, J., & Xiao, Y. (2021). Misinformation about COVID-19: Evidence for differential latent profiles and a strong association with trust in science. *BMC Public Health*, *1*(89), 1–12. https://bmcpublichealth.biomedcentral.com/counter/pdf/10.1186/s12889-020-10103-x.pdf. doi:10.118612889-020-10103-x PMID:33413219

Ahmad (2020, May 9). Nigeria appreciates Chinese support in COVID-19 fight. *China Daily*. https://global.chinadaily.com.cn/a/202005/09/WS5eb5fed9a310a8b241154601.html

Ahmad, A. R., & Murad, H. R. (2020). The Impact of Social Media on Panic During the COVID-19 Pandemic in Iraqi Kurdistan: Online Questionnaire Study. *Journal of Medical Internet Research*, *22*(5), e19556. doi:10.2196/19556 PMID:32369026

Akaydın Aydın, A. ve Sadakaoğlu, M. C. (2020). İdeoloji ve instagram: Türk silahlı kuvvetlerinin yönettiği ınstagram hesabının kamu diplomasisi açısından incelenmesi. *The Turkish Online Journal of Design. Art and Communication*, *10*(3), 221–231.

Aksoy, H. (2021). *Excursus: Turkey's military engagement abroad.* Retrieved August 20, 2021, from https://www.cats-network.eu/topics/visualizing-turkeys-foreign-policy-activism/excursus-turkeys-military-engagement-abroad#c4312

Alanagreh, L., Alzoughool, F., & Atoum, M. (2020). The human coronavirus disease COVID-19: Its origin, characteristics, and insights into potential drugs and its mechanisms. *Pathogens (Basel, Switzerland)*, *9*(5), 331. https://www.ncbi.nlm.nih.gov/pmc/articles/PMC7280997/. doi:10.3390/pathogens9050331 PMID:32365466

Al-Herz, W., Bousfiha, A., Casanova, J. L., Chatila, T., Conley, M. E., Cunningham-Rundles, C., Etzioni, A., Franco, J. L., Gaspar, H. B., Holland, S. M., & Klein, C. (2014, April 22). Primary immunodeficiency diseases: An update on the classification from the international union of immunological societies expert committee for primary immunodeficiency. *Frontiers in Immunology*, *5*, 162. PMID:24795713

Allam, Z. & Jones, D. S. (2020). On the coronavirus (COVID-19) outbreak and the smart city network: universal data sharing standards coupled with artificial intelligence (AI) to benefit urban health monitoring and management. In *Healthcare, 8*(1), p. 46. Multidisciplinary Digital Publishing Institute.

Ambassador. (2020). *Ambassador Zhou Pingjian's Exclusive Interview with Punch.* Embassy of the People's Republic of China in the Federal Republic of Nigeria. http://ng.china-embassy.gov.cn/eng/zngx/cne/202005/t20200507_7775332.htm

Andreas, S. 2012. Social Media diplomacy: the rules of engagement. *Diplo*. https://www.diplomacy.edu/blog/social-media-diplomacy-rules-engagement

Antwi-Boateng, O., & Al Mazrouei, K. A. M. (2021). The challenges of digital diplomacy in the era of globalization: The case of the United Arab Emirates. *International Journal of Communication*, *15*(1), 4577–4595.

Argentino, M. A., & Amarasingam, A. (2020). The COVID conspiracy files. A Technical report. New York: GNET: Global Network on Extremism and Technology.

Atlani-Duault, L., Ward, J., Roy, M., Morin, C., & Wilson, A. (2020). Tracking online heroisation and blame in epidemics. *The Lancet. Public Health*, *5*(3), e137–e138. doi:10.1016/S2468-2667(20)30033-5 PMID:32085818

Awoniyi, S. (2015). African cultural values: The past, present and future. *Journal of Sustainable Development in Africa*, *17*(1), 1–13.

Babalola, T. A. (2019). Decolonising digital humanities: The African perspective. In L. Elisabeth & J. Wernimontr (Eds.), *Bodies of information: International feminism and digital humanists* (pp. 307–313). University of Minnesota Press.

Babu, B. S., Kamalakannan, J., Meenatchi, N., Karthik, S., & Boopathi, S. (2022). Economic impacts and reliability evaluation of battery by adopting Electric Vehicle. *IEEE Explore*, 1–6.

Baker, T. (2016). Bringing the human back to work. *HR Daily Community*. https://community. hrdaily.com.au/profiles/blogs/bringing-the-human-being-back-to-work

Balcı, A. (2018). Savunma Diplomasisi Kavramı, Özellikleri ve Uygulamaları. *Ufuk Üniversitesi Sosyal Bilimler Enstitüsü Dergisi*, *7*(14), 45–58.

Barometre, A. (2020). *Africans' perception about China: A sneak peek from 18 countries.* Afro Barometre.

BBC News. (2020). Coronavirus: Chinese chief for Kano say make pipo no fear dem. *BBC News Pidgin.* https://www.bbc.com/pidgin/tori-51702073

BBC News. (2021). Covid origin: Why is the Wuhan lab-leak theory taken seriously? *BBC News* https://www.bbc.com/news/world-asia-china-57268111

Bello, A., Wiebe, N., Garg, A., & Tonelli, M. (2015). Evidence-based decision-making: Systematic reviews and meta-analysis. *Methods in Molecular Biology (Clifton, N.J.)*, *12*(1), 397–416. doi:10.1007/978-1-4939-2428-8_24 PMID:25694324

Benoit, W. L. (2015). *Image restoration theory*. Wiley Online Library/ https://doi. org/10.1002/9781405186407.wbieci009.pub2

Binks, A. P., LeClair, R. J., Willey, J. M., Brenner, J. M., Pickering, J. D., Moore, J. S., Huggett, K. N., Everling, K. M., Arnott, J. A., Croniger, C. M., Zehle, C. H., Kranea, N. K., & Schwartzstein, R. M. (2021, February 18). Changing Medical Education, Overnight: The Curricular Response to COVID-19 of Nine Medical Schools. *Teaching and Learning in Medicine*, *33*(3), 1–9. doi:1 0.1080/10401334.2021.1891543 PMID:33706632

Bjola, C. (2017). *Trends and counter-trends in digital diplomacy. Working paper no18. "Digital Diplomacy in the 21ˢᵗ Century" project*. Munich: German Ministry of Foreign Affairs.

Bjola, C., & Manor, I. (2020). *Digital diplomacy in the time of the coronavirus pandemic*. USC Public Diplomacy. https://uscpublicdiplomacy.org/blog/digital-diplomacy-time-coronavirus-pandemic

Bjola, C. (2018). *Diplomacy in the digital age*. Real Institute/ Elcano.

Bjola, C. (2019). *Diplomacy in the age of artificial intelligence*. The Emirate Diplomatic Academy.

Bjola, C. (2022). Digital diplomacy as world disclosure: The case of the COVID-19 pandemic. *Place Branding and Public Diplomacy*, *18*(2), 22–25. doi:10.105741254-021-00242-2

Bjola, C., & Manor, I. (2022). The rise of hybrid diplomacy: From digital adaptation to digital adoption. *International Affairs*, *98*(2), 471–491. doi:10.1093/ia/iiac005

Bodomo, A. (2018). Is China colonising Africa? In S. Raudino & A. Poletti (Eds.), *Global economic governance and human development* (pp. 122–135). Routledge. doi:10.4324/9781315169767-7

Bolsen, T., Palm, R., & Kingsland, J. T. (2020). Framing the Origins of COVID-19. *Science Communication*, *42*(5), 562–585. https://doi.org/10.1177/1075547020953603

Boopathi, S., Arigela, S. H., Raman, R., Indhumathi, C., Kavitha, V., & Bhatt, B. C. (2022). Prominent Rule Control-based Internet of Things: Poultry Farm Management System. *IEEE Explore*, 1–6.

Boopathi, S., Siva Kumar, P. K., & Meena, R. S. J., S. I., P., S. K., & Sudhakar, M. (2023). Sustainable Developments of Modern Soil-Less Agro-Cultivation Systems. In Human Agro-Energy Optimization for Business and Industry (pp. 69–87). IGI Global. https://doi.org/ doi:10.4018/978-1-6684-4118-3.ch004

Brem, A., Viardot, E., & Nylund, P. A. (2021, February 1). Implications of the coronavirus (COVID-19) outbreak for innovation: Which technologies will improve our lives? *Technological Forecasting and Social Change*, *163*, 120451. doi:10.1016/j.techfore.2020.120451 PMID:33191956

Budd, J., Miller, B. S., Manning, E. M., Lampos, V., Zhuang, M., Edelstein, M., Rees, G., Emery, V. C., Stevens, M. M., Keegan, N., Short, M. J., Pillay, D., Manley, E., Cox, I. J., Heymann, D., Johnson, A. M., & McKendry, R. A. (2020). Digital technologies in the public-health response to COVID-19. *Nature Medicine*, *26*(8), 1183–1192. doi:10 103841591-020-1011-4 PMID:32770165

Burnard, M., & Richards, A. (2020). COVID-19 and 5G: Biggest cover-up in history? True or false? *INcontext International*. https://www.incontextinternational.org/2020/04/02/covid-19-and-5gbiggest-cover-up-in-history-true-or-false/

Campbell, J. (2020). Despite new China-Africa tension, Beijing has pivotal role to play in Africa's COVID-19 recovery. *Foreign Relations*. https://www.cfr.org/blog/despite-new-china-africa-tension-beijing-has-pivotal-role-play-africas-covid-19-recovery

Carvalho, G. (2020, June 11). *Africa must unmute its mic as e-diplomacy takes root*. ISS Africa. https:// issafrica.org/iss-today/africa-must-unmute-its-mic-as-e-diplomacy-takes-root

Chang, C. K., & Fung, A. Y. H. (2021). From soft power to sharp power: China's image in Hong-Kong health crises from 2003-2020. *Global Media and China*, *6*(1), 62–76. doi:10.1177/2059436420980475

Chauhan, A., Jakhar, S. K., & Chauhan, C. (2021, January 10). The interplay of circular economy with industry 4.0 enabled smart city drivers of healthcare waste disposal. *Journal of Cleaner Production*, *279*, 123854. doi:10.1016/j.jclepro.2020.123854 PMID:32863607

Chinese Embassy. (2020). *Fighting Covid-19 China in Action*. Embassy of the People's Republic of China in the Federal Republic of Nigeria. http://ng.china-embassy.gov.cn/eng/zngx/cne/202006/t20200608_7775354.htm

Chinese Embassy. (2020a). *Statement by Press Secretary of the Embassy of China in Nigeria.* Embassy of the People's Republic of China in the Federal Republic of Nigeria. http://ng.china-embassy.gov.cn/eng/zngx/cne/202004/t20200430_7775316.htm

Chinese Embassy. (2020b). *Statement by Press Secretary of the Embassy of China in Nigeria.* Embassy of the People's Republic of China in the Federal Republic of Nigeria. http://ng.china-embassy.gov.cn/eng/zngx/cne/202007/t20200708_7775390.htm

Chow, D. C. K. (2000). Counterfeiting in the People's Republic of China. *Washington University Law Quarterly*, *78*(1). https://openscholarship.wustl.edu/law_lawreview/vol78/iss1/1

Christensen, T. J. (2020). *A modern tragedy? COVID-19 and US-China relations.* Foreign Policy.

Chu, M. (2020a). China will defeat the Coronavirus. Vanguard.

Chu, M. (2020b). Nigeria's support helpful to fight Coronavirus. Vanguard.

Cinelli, M., Quattrociocchi, W., Galeazzi, A., Valensise, C. M., Brugnoli, E., Schmidt, A. L., Zola, P., Zollo, F., & Scala, A. (2020). The COVID-19 social media infodemic. *Scientific Reports*, *10*(1), 16598. doi:10.103841598-020-73510-5 PMID:33024152

Cooper, A. F., Heine, J., & Thakur, R. (2013). Introduction: The challenges of 21st-century diplomacy. In A. F. Cooper, J. Heine, & R. Thakur (Eds.), *The Oxford Handbook of Modern Diplomacy* (pp. 2–27). Oxford University Press. doi:10.1093/oxfordhb/9780199588862.001.0001

Copeland, D. (2012) Public diplomacy, branding, and the image of nations, part iv: Some practical implications. *USC Center on Public Diplomacy.* https://uscpublicdiplomacy.org/blog/public-diplomacy-branding-and-image-nations-part-iv-some-practical-implications

Cottey, A., & Forster, A. (2004). *Reshaping defense diplomacy: New roles for military cooperation and assistance.* Oxford University Press.

Cull, N. J. (2008). Public diplomacy: Taxonomies and histories. *The Annals of the American Academy of Political and Social Science*, *616*(1), 31–54. doi:10.1177/0002716207311952

Cull, N. J. (2009). *Public diplomacy: Lessons from the past.* Figueora Press.

Daily Trust. (2020). Editorial: Maltreatment of Nigerians in China. *Daily Trust.*

Daxue Consulting. (2020). *The AI ecosystem in China.* Daxue Consulting.

De Coninck, D., Frissen, T., Matthijs, K., d'Haenens, L., Lits, G., Champagne-Poirier, O., Carignan, M., David, M., Pignard-Cheynel, N., Salerno, S., & Généreux, M. (2021). Beliefs in conspiracy theories and misinformation about COVID-19: comparative perspectives on the role of anxiety, depression, and exposure to and trust in information sources. *Frontier Psychology*, 1-13. doi:10.3389/fpsyg.2021.646394

Debata, B., Patnaik, P., & Mishra, A. (2020, November). COVID-19 pandemic! It's impact on people, economy, and environment. *Journal of Public Affairs*, *20*(4), e2372. doi:10.1002/pa.2372

Compilation of References

Defense Academy of the United Kingdom. (2009). *A History Of RCDS*. Retrieved July 27, 2022, from https://webarchive.nationalarchives.gov.uk/ukgwa/20091211061225/http://www.da.mod.uk/colleges/rcds/About_Us/A%20History%20of%20RCDS

Department of the Army. (2001). *US Army Field Manual 3-0, Operations*. United States Army.

Depoux, A., Martin, S., Karafillakis, E., Preet, R., Wilder-Smith, A., & Larson, H. (2020). The pandemic of social media panic travels faster than the COVID-19 outbreak. *Journal of Travel Medicine*, *27*(3), taaa031. doi:10.1093/jtm/taaa031 PMID:32125413

Di Martino, L. (2020). Conceptualising Public Diplomacy Listening on Social Media. *Place Branding and Public Diplomacy*, *16*(2), 131–142. doi:10.105741254-019-00135-5

Digital Diplomacy Blog. (2018). How diplomat can combat digital propaganda. *Exploring digital diplomacy*. https://digdipblog.com/2018/06/25/how-diplomats-can-combat-digital-propaganda/

Dikotter, F. (1992). *The discourse of race in modern china* (Vol. C). Hurst and Co Publishers Ltd.

Dilshan, N. W. T. (2020). *A Workforce Perception Of The Impact Of Covid-19 (Novel Coronavirus) On Job Security Of Tourism Industry In Sri Lanka*. UOC Tourism.

Diplomacy Data. (2016). History of digital diplomacy and main milestones. *Diplomacy Data*. http://diplomacydata.com/history-of-digital-diplomacy-and-main-milestones/

Dizrad, W. Jr. (2001). *Digital diplomacy: US foreign policy in the information age*. Praeger.

Doğan, H. (2014). Türk silahlı kuvvetlerinin kamu diplomasisi faaliyetleri. *Güvenlik Bilimleri Dergisi*, *3*(2), 67–90. doi:10.28956/gbd.283039

Domakonda, V. K., Farooq, S., Chinthamreddy, S., Puviarasi, R., Sudhakar, M., & Boopathi, S. (2022). Sustainable Developments of Hybrid Floating Solar Power Plants: Photovoltaic System. In Human Agro-Energy Optimization for Business and Industry (pp. 148–167). IGI Global.

Dong, N. (2017). 'Unequal Sino-African Relationships': A Perspective from Africans in Guangzhou. In Y.-C. Kim (Ed.), *China and Africa: A New Paradigm of Global Business* (pp. 237–259). Palgrave Macmillan.

Eggeling, K. A., & Nissen, R. (2021). The synthetic situation in diplomacy: Scopic media and the digital mediation of estrangement. *Global Studies Quarterly*, *1*(2), 102–119. doi:10.1093/isagsq/ksab005

Ehanire, O. (2020, March 26). Coronavirus: Nigeria receives donations from China to fight Covid-19. *Premiumtimes*. https://www.premiumtimesng.com/news/more-news/384094-coronavirus-nigeria-receives-donations-from-china-to-fight-covid-19.html

Endong, F. P. C. (2020). Digitization of African public diplomacy: Issues, challenges, and opportunities. *International Journal of Digital Society*, *11*(2), 1607–1618. doi:10.20533/ijds.2040.2570.2020.0201

Endong, F. P. C. (2022). Re-branding China's battered image in Nigeria amidst the COVID-19 pandemic. A qualitative analysis of Chinese diplomatic communications. *Journal of BRICS Studies*, *1*(1), 26–40. doi:10.36615/jbs.v1i1.615

Erdoğan, A. (2022). *Savunma diplomasisi ve milli güç*. Retrieved March 2 2022, from https://thinktech.stm.com.tr/tr/turk-savunma-sanayiinin-adaptasyon-ve-donusumunde-kuresel-oyuncularla-rekabet

Erenel, F. (2020). *Türkiye'nin savunma diplomasisi ve yurtdışında asker bulundurma*. Retrieved December 8, 2020, from https://www.gazetebirlik.com/yazarlar/turkiyenin-savunma-diplomasisi-ve-yurt-disinda-askeri-varlik-bulundurma/

European Council on Foreign Relations. (2021). *Turkey's drone diplomacy: Lessons for Europe*. Retrieved January 31, 2022, from https://ecfr.eu/article/turkeys-drone-diplomacy-lessons-for-europe/

Ezekwesili, O. (2020, April 17). China must pay reparations to Africa for its coronavirus failures. *The Cable*. https://www.thecable.ng/china-must-pay-reparations-to-africa-for-its-coronavirus-failures

Faye, M. (2000). *Developing national information and communication infrastructure policies and plans in Africa*. Paper presented at the Nigeria NICI workshop, Abuja, Nigeria.

Fleming, D. C. (2014). Counterfeiting in China. *East Asia Law Review*, *10*, 14.

FOCAC. (2020). *Joint Statement of the Extraordinary China-Africa Summit On Solidarity Against COVID-19*. Embassy of the People's Republic of China in the Federal Republic of Nigeria. http://ng.china-embassy.gov.cn/eng/zngx/cne/202006/t20200618_7775366.htm

Foreign Affairs Office. (2020). *One world, one fight in solidarity we stand for the building of a community of common health for mankind*. Embassy of the People's Republic of China in the Federal Republic of Nigeria. http://ng.china-embassy.gov.cn/eng/zngx/cne/202004/t20200426_7775307.htm

Foreign Ministry. (2020). *Foreign Ministry spokesperson: We have zero tolerance for discrimination*. Embassy of the People's Republic of China in the Federal Republic of Nigeria. http://ng.china-embassy.gov.cn/eng/zngx/cne/202004/t20200411_7775286.htm

Galán, M., Velasco, M., Casas, M., & Goyanes, M. (2020). *SARS-CoV-2 Seroprevalence Among All Workers in a Teaching Hospital in Spain: Unmasking The Risk*. Cold Spring Harbor Laboratory.

Gambhir, M. (2021). *Defense Diplomacy and its Relevance*. Retrieved July 28, 2022, from https://www.claws.in/defense-diplomacy-and-its-relevance/

Gauttam, P., Singh, B., & Kaur, J. (2020). COVID-19 and Chinese Global Health Diplomacy: Geopolitical Opportunity for China's Hegemony? *Millennial Asia*, *11*(3), 318–340. doi:10.1177/0976399620959771

Geerts, S., Xunwa, N., & Rossouw, D. (2014). *Africans' perceptions of Chinese business in Africa. A survey (Globethics net Focus No. 18)*. Ethics Institute of South Africa.

George, B. & Paul, J. (2020). *Digital Transformation in Business and Society Theory and Cases: Theory and Cases.* Springer. . doi:10.1007/978-3-030-08277-2

German Institute. (2018). *New realities in foreign affairs: Diplomacy in the 21st Century.* German Institute for International and Security Affairs. https://www.swp-berlin.org/en/publication/new-realities-in-foreign-affairs-diplomacy-in-the-21st-century

Ghose, S. (2020). *Crisis as catalyst: The COVID-19 impact on innovation.* Berkeley. Sutardia Center or Entrepreneurship & Technology.

Gluck, A. (2015). *De-Westernisation. A key concept paper.* University of Leeds.

González-Padilla, D. A., & Tortolero-Blanco, L. (2020). *Social media influence in the COVID-19 Pandemic.* National Institutes of Health., doi:10.1590/S1677-5538.IBJU.2020.S121

Gordon, L. (2020). Humanism: Africa. *Encyclopaedia of African philosophy.*

Grattan, R. F. (2011). *Strategic Review: The Process of Strategy Formulation in Complex Organisations.* Gower Publishing.

Grina, G. (2017). National Military Diplomacy and its Prospects. *Lithuanian Annual Strategic Review*, *15*(1), 153–177. doi:10.1515/lasr-2017-0007

Grincheva, N. (2021). Digital diplomacy in the midst of (post) pandemic crisis: Inspiring, educating, and contributing to global peace and well-being. *MW21.* https://mw21.museweb.net/paper/digital-diplomacy-in-the-midst-of-post-pandemic-crisis-inspiring-educating-contributing-to-global-peace-and-well-being/

Guangzhou. (2020). *Guangzhou: Facts, solidarity and cooperation.* Embassy of the People's Republic of China in the Federal Republic of Nigeria. http://ng.china-embassy.gov.cn/eng/zngx/cne/202005/t20200504_7775320.htmJidda

Gu, F., Wu, Y., Hu, X., Guo, J., Yang, X., & Zhao, X. (2021). The role of conspiracy theories in the spread of COVID-19 across the United States. *International Journal of Environmental Research and Public Health*, *18*(7), 1–14. doi:10.3390/ijerph18073843 PMID:33917575

Gupta, A. (2020). Clashes over COVID-19 aid in Nigeria. China Africa Project. *China African Project.* https://chinaafricaproject.com/student-xchange/clashes-over-covid-19-aid-in-nigeria/

Haber, T. R. T. (2022). *Türk teknoloji şirketinin askeri metaverse uygulaması ilk kez NATO'da tanıtıldı.* Retrieved August 14, 2022, from https://www.trthaber.com/haber/bilim-teknoloji/turk-teknoloji-sirketinin-askeri-metaverse-uygulamasi-ilk-kez-natoda-tanitildi-701333.html

Haleem, A., & Javaid, M. (2020, December 1). Medical 4.0 and its role in healthcare during COVID-19 pandemic: A review. *Journal of Industrial Integration and Management.*, *5*(4), 531–545. doi:10.1142/S2424862220300045

Haslam, N. (2006). Dehumanization: An integrative review. *Personality and Psychology Review*, *10*(3), 252–264. doi:10.120715327957pspr1003_4 PMID:16859440

Hayden, C. (2012). Social media at stake: Power, practice, and conceptual limits for US public diplomacy. *Global Media*, *25*, 1–15.

Hedling, E., & Bremberg, N. (2021). Practice approaches to the digital transformations of diplomacy: Toward a new research agenda. *International Studies Review*, *1*(1), 1–24. doi:10.1093/isr/viab027

Hocking, B., & Melissen, J. (2015, July 12). Diplomacy in the digital age. *Clingendael Magazine*. https://www.clingendael.org/sites/default/files/pdfs/Digital_Diplomacy_in_the_Digital%20 Age_Clingendael.pdf

House of Commons. (1998). *The Strategic Defense Review White Paper*. Retrieved 28.07.2022 from https://researchbriefings.files.parliament.uk/documents/RP98-91/RP98-91.pdf

Human Rights Watch. (2020). *China: COVID-19 Discrimination against Africans: Forced quarantine, evictions and refused services in Guangzhou*. Human Right Watch.

Hung, K. (2015). *Repairing the "made-in-China" image in the U.S. and U.K.: effects of Government-supported advertising. International Public Relations And Public Diplomacy*. Peter Lang Publishing, Inc.

Idang, G. E. (2015). African culture and values. *Phronomon*, *16*(2), 97–111. doi:10.25159/2413-3086/3820

IFRC. (2020). *COVID-19 two years on: A new normal for some while millions still at risk, warns Red Cross Red Crescent*. IFRC.

Ihekweazu, C. (2020, March 4). Coronavirus: NCDC DG Chikwe Ihekweazu is sharing what he learned on his trip to China. *Bellanaija*. https://www.bellanaija.com/2020/03/coronavirus-chikwe-ihekweazu-china/

Ittefaq, M. (2019). Digital diplomacy via social networks: A cross-national analysis of governmental usage of Facebook and Twitter for digital engagement. *Journal of Contemporary Eastern Asia*, *18*(1), 49–69.

Jackson, S.F. (2019). Two distant giants. China and Nigeria perceive each other. *European Middle Eastern and African Affairs*, 40-74.

Javaid, M., Haleem, A., Singh, R. P., Haq, M. I., Raina, A., & Suman, R. (2020, December 1). Industry 5.0: Potential applications in COVID-19. *Journal of Industrial Integration and Management.*, *5*(4), 507–530. doi:10.1142/S2424862220500220

Javaid, M., Haleem, A., Vaishya, R., Bahl, S., Suman, R., & Vaish, A. (2020, July 1). Industry 4.0 technologies and their applications in fighting COVID-19 pandemic. *Diabetes & Metabolic Syndrome*, *14*(4), 419–422. doi:10.1016/j.dsx.2020.04.032 PMID:32344370

Jayatilaka, C. (2020) The Effects of Digital Diplomacy on International Relations: Lessons for Sri Lanka https://lki.lk/publication/the-effects-of-digital-diplomacy-on-international-relations-a-lesson-for-sri-lanka/

Jeevanantham, Y. A., Saravanan, A., Vanitha, V., Boopathi, S., & Kumar, D. P. (2022). Implementation of Internet-of Things (IoT) in Soil Irrigation System. *IEEE Explore*, 1–5.

Jiang, Y. (2021). Problematic Social Media Usage and Anxiety Among University Students During the COVID-19 Pandemic: The Mediating Role of Psychological Capital and the Moderating Role of Academic Burnout. *Frontiers in Psychology*, *12*, 612007. doi:10.3389/fpsyg.2021.612007 PMID:33613391

Jia, W., & Lu, F. (2021). US media's coverage of China's handling of COVID-19: Playing the role of the fourth branch of government or the fourth estate? *Global Media and China*, *6*(1), 62–76. doi:10.1177/2059436421994003

Jinping, X. (2020). *Keynote Speech by H.E. Xi Jinping President of the People's Republic of China at the Extraordinary China-Africa Summit on Solidarity against COVID-19*. Embassy of the People's Republic of China in the Federal Republic of Nigeria. http://ng.china-embassy.gov.cn/eng/zngx/cne/202006/t20200617_7775363.htm

Joshi, A. M., Shukla, U. P., & Mohanty, S. P. (2008). *Smart Healthcare for Diabetes: A COVID-19 Perspective*. arXiv preprint arXiv:2008.11153.

Joshi, A. (2021, January 6). COVID-19 pandemic in India: Through psycho-social lens. *Journal of Social and Economic Development*, *23*(S2), 1–24. doi:10.100740847-020-00136-8 PMID:34720479

Kadam, A., & Atre, S. (2020). Social media panic and COVID-19 in India. *Journal of Travel Medicine*, *27*. doi:10.1093/jtm/taaa057 PMID:32307545

Kaiser, M. S., Al Mamun, S., Mahmud, M., & Tania, M. H. (2021). *Healthcare robots to combat COVID-19. InCOVID-19: Prediction, Decision-Making, and Its Impacts*. Springer.

Kehbuma, L. (2016). *Globalisation and cyberculture. An Afrocentric perspective*. Springer.

Kerry, J. (2013). Digital Diplomacy: Adapting Our Diplomatic Engagement. *DipNote*. U.S Department of State Official Blog, 6/V/2013. http://2007-2017-blogs.state.gov/stories/2013/05/06/digital-diplomacyadapting-our-diplomatic-engagement.html

Khan, H., Kushwah, K.K., & Singh, S. (2021). Smart technologies driven approaches to tackle COVID-19 pandemic: a review. *Biotech*, *3*(11), 50. doi:10.1007/s13205-020-02581-y

Khan, K. S., Kunz, R., Kleijnen, J., & Antes, G. (2003). Five steps to conducting a systematic review. *Journal of the Royal Society of Medicine*, *96*(3), 118–121. doi:10.1177/014107680309600304 PMID:12612111

Khern, N. C. (2009). On command. *Pointer: Journal of the Singapore Armed Forces Supplement*, *34*(1), 1–28.

Kobus, J., & Bryan, R. (2018). *China's impact on the African renaissance: the baobab grows*. Palgrave Macmillan.

Ko, C.-H., Yen, C.-F., Yen, J.-Y., & Yang, M.-J. (2006). Psychosocial impact among the public of the severe acute respiratory syndrome epidemic in Taiwan. *Psychiatry and Clinical Neurosciences*, *60*(4), 397–403. doi:10.1111/j.1440-1819.2006.01522.x PMID:16884438

Koerner, W. (2006). Security sector reform: Defense diplomacy. *Library of Parliament. Parliamentary Information and Research Services Brief*, *6*(12), 1–3.

Kolawole, Y. (2022, February 21). Trade deficit with China worsens with $23 bn imports in 2021. *Vanguard*. https://www.vanguardngr.com/2022/02/trade-deficit-with-china-worsens-with-23bn-imports-in-2021/

Krishnan, A. (2020). The COVID-19 pandemic is China's biggest crisis since Tiananmen, says Richard McGregor. *The Hindu*. https://www.thehindu.com/news/national/coronavirus-the-covid-19-pandemic-is-chinas-biggest-crisis-since-tiananmen-says-richard-mcgregor/article31438204.ece

Kumara, V., Mohanaprakash, T. A., Fairooz, S., Jamal, K., Babu, T., & B., S. (2023). Experimental Study on a Reliable Smart Hydroponics System. In *Human Agro-Energy Optimization for Business and Industry* (pp. 27–45). IGI Global. https://doi.org/ doi:10.4018/978-1-6684-4118-3.ch002

Kurbalija, J. (2020). Negotiating with robot: Future or fiction? *Diplo*. https://www.diplomacy.edu/blog/negotiating-robots-future-or-fiction

Kurbalija, J. (2022). Digital diplomacy. *Diplo*. https://www.diplomacy.edu/topics/digital-diplomacy/

Kurbalija, J. (2017). The impact of the internet and ICT on contemporary diplomacy. In P. Kerr (Ed.), *Diplomacy in a globalising world* (pp. 151–169). Oxford University Press.

Labott, E. (2022). *Redefining diplomacy in the wake of the COVID-19 pandemic*. The Meridian Center for Diplomatic Engagement.

Lamsal, H. L. (2022). Effectiveness of military diplomacy towards nationalism, national security, and unity. *Unity Journal*, *3*(01), 82–96. doi:10.3126/unityj.v3i01.43317

Lancet. (2020). COVID-19: Fighting panic with information. *Lancet*, *395*(10224), 537. doi:10.1016/S0140-6736(20)30379-2 PMID:32087777

Lang, B. (2019). *China and global integrity building: Challenges and prospects for engagement*. Michelsen Institute (CMI).

Lan, S. (2016). The Shifting Meanings of Race in China: A Case Study of the African Diaspora Communities in Guangzhou. *City & Society*, *28*, 298–318. doi:10.1111/ciso.12094

Laverty, A. (2012). *The norms of African diplomatic culture: Implication for African integration*. London: IR 503

Lemon, E., & Jardine, B. (2021). Central Asia's multi-vector diplomacy. *Kennan Cable*, (68), 1–12.

Lijian, Z. (2020a). *Foreign Ministry Spokesperson Zhao Lijian's Remarks on Guangdong's anti-epidemic measures concerning African citizens in China*. Embassy of the People's Republic of China in the Federal Republic of Nigeria. http://ng.china-embassy.gov.cn/eng/zngx/cne/202004/t20200413_7775302.htm

Lijian, Z. (2020b). *Foreign Ministry spokesperson Zhao Lijian's remarks on Nigerian Foreign Minister Onyeama's address to the press*. Embassy of the People's Republic of China in the Federal Republic of Nigeria. http://ng.china-embassy.gov.cn/eng/zngx/cne/202004/t20200418_7775304.htm

Lijian, Z. (2020c). *Embassy spokesperson's remarks on national security legislation for Hong Kong SAR*. Embassy of the People's Republic of China in the Federal Republic of Nigeria. http://ng.china-embassy.gov.cn/eng/zngx/cne/202005/t20200522_7775339.htm

Linvingstone, S. D. (2011). *Less than human: Why we demean, enslave and exterminate other*. St. Martin's Press.

Liu, L., Huang, Y., & Jin, J. (2022). China's vaccine diplomacy and its implications for global health governance. *Health Care*, *10*(7), 1276. doi:10.3390/healthcare10071276

Losifidis, P., & Wheeler, M. (2016, May 13). *Public diplomacy 2.0 and social media*. Research Gate. https://www.researchgate.net/publication/303097870_public_diplomacy_20_and_the_social_media

Louw, R. (2010). Africanisation: A rich environment for active learning on a global platform. *Progressio*, *32*(1), 42–54.

Lowy Interpreter. (2015, April 15). Does India do digital diplomacy? *Lowy Interpreter*. http://www.lowyinterpreter.org

Madan Gopal, K. (2020) Importance Of Civil Society Organisations In Managing Covid-19 Pandemic. *Outlook India*. https://www.outlookindia.com/website/story/india-news-the-importance-of-the-role-of-civil-society-organisations-in-managing-the-pandemic/364223

Madrid-Morales, D. (2017). China's digital public diplomacy towards Africa: Actor, messages and audiences, 129-146. Routledge.

Manor, I. (2018). *The digitalisation of diplomacy: Towards clarification of a fractured terminology. Working Papers Series*. Oxford: Oxford Digital Diplomacy Research Group.

Manor, I. (2021). The Digital Legacy of Covid-19 https://www.e-ir.info/2021/04/02/the-digital-legacy-of-covid-19/

Manor, I. (2018). *The Digitization of public diplomacy*. Palgrave.

Manor, I. (2019). Digital diplomacy in Africa: A research agenda. *Hague Journal of Diplomacy*, *10*(4), 538–574.

Masolo, D. A. (2018). *Humanity: African thought. Encyclopedia of Science and Philosophy. Heterodyne to Hydrazoic Acid.* Science Rank.

Masters, L. (2021). Africa, the Fourth Industrial Revolution, and digital diplomacy: (Re)Negotiating the international knowledge structure. *South African Journal of International Affairs, 28*(3), 361–377. doi:10.1080/10220461.2021.1961605

McCourt, D. M. (2016). Practice theory and relationalism as the new constructivism. *International Studies Quarterly, 60*(3), 475–485. doi:10.1093/isqqw036

McGinnis, D. (2020, October 27). What is the fourth industrial revolution. *Sales Force.* https://www.salesforce.com/blog/what-is-the-fourth-industrial-revolution-4ir

Mearsheimer, J. (1990). Back to the future: Instability in Europe after the cold war. *International Security, 15*(1), 5–56. doi:10.2307/2538981

Melissen, J. (2015). *The new public diplomacy: Soft power in international relations.* Palgrave Macmillan.

Melissen, J., & de Keulenaar, E. (2017). Critical digital diplomacy as a global challenge: The South Korean experience. *Global Policy, 8*(3), 294–302. doi:10.1111/1758-5899.12425

Mertha, C. (2005). The politics of piracy: Intellectual property in contemporary China. *Law & Society Review, 40*(4).

Metin, M. (2022). *Turkish defense industry gains its interdependence.* Retrieved August 16, 2022, from https://businessdiplomacy.net/turkish-defense-industry-gains-its-independence/

Metzger, E. T. (2012). Is it the medium or the message? Social media, American public relations and Iran. *Global Media Journal, 38*, 1–16.

Mhlanga, D., & Moloi, T. (2020). Covid-19 and the digital transformation of education: What are we learning on 4IR in South Africa? *Education Sciences, 10*(2), 1–11. doi:10.3390/educsci10070180

Midhio, I. W., & Priyono, J. (2019). Education and research as components of Indonesia's defense diplomacy. *Jurnal Pertahanan, 5*(1), 61–70. doi:10.33172/jp.v5i1.487

Miller, D. C. (2015). *SIMNET and Beyond: A History of the Development of Distributed Simulation.* Retrieved July 29, 2022, from https://www.iitsec.org/-/media/sites/iitsec/link-attachments/iitsec-fellows/2015_fellowpaper_miller.ashx

Milli Savunma Bakanlığı. (2021). *2020 yılı faaliyet raporu.* Bütçe ve Mali Hizmetler Genel Müdürlüğü.

Milli Savunma Bakanlığı. (2022). *Yurtdışındaki askeri ateşelikler.* Retrieved August 15, 2022, from https://www.msb.gov.tr/SavunmaGuvenlik/icerik/yurt-disindaki-asker-ataselikler

Mitchell, A., & Oliphan, J. (2020 March 18). Americans Immersed in COVID-19 News; Most Think Media Are Doing Fairly Well Covering It. *Pew Research Center* [Blog Post]. https://www.journalism.org/2020/03/18/americans-immersed-in-covid-19-news-mostthink-media-are-doing-fairly-well-covering-it/

Moll, I. (2022). The fourth industrial revolution: A new ideology. *TripleC*, *20*(1), 45–61. doi:10.31269/triplec.v20i1.1297

Moumita de das. (2020). *Digital diplomacy: the trend is here to stay*. Adamas University. HTTPS://adamasuniversity.ac.in/digital-diplomacy-the-trend-is-here-to-stay.

Mujawar MA, Gohel H, Bhardwaj SK, Srinivasan S, Hickman N, & Kaushik A. (2020). Aspects of nano-enabling biosensing systems for intelligent healthcare; towards COVID-19 management. *Materials Today Chemistry*.

Mullard, S., & Aarvik, P. (2020). Supporting civil society during the Covid-19 pandemic. The potentials of online collaborations for social accountability. *E-International Relations*. https://www.u4.no/publications/supporting-civil-society-during-the-covid-19-pandemic

Muller, S., Brazys, S., & Dukalskis, A. (2021). *Discourse wars and Mask Diplomacy: China's global image management in times of crisis (Working Paper 109)*, Dublin: AIDDATA.

Mulualem, M. (2014, April 26). America is coming to Africa: a comparison with China. *AllAfrica*. http://allafrica.com/stories/201404280729.html?viewall=1

Muniruzzaman, A. N. M. (2020). Defense diplomacy: A powerful tool of statecraft. *CLAWS Journal*, *13*(2), 63–80.

Murithi, T. (2006). African approaches to building peace and social solidarity. *Accord: African Journal of Community Relations*, *2*, 1–14.

Nasiru, J. (2020). Video: Stop seizing the passports of our citizens – Nigerian diplomat tackles Chinese officials. *The Cable*. https://www.thecable.ng/video-stop-seizing-the-passports-of-our-citizens-nigerian-diplomat-tackles-chinese-officials

Natarajan, K. (2014). Digital public diplomacy and a strategic narrative for India. *Strategic Analysis*, *38*(1), 91–106. doi:10.1080/09700161.2014 863478

Nathan, J. (1993). Force, order, and diplomacy in the age of Louis XIV. *The Virginia Quarterly Review*, *69*(4), 633–649.

Nato. (2020). *Partnership for Peace programme*. Retrieved July 28, 2022, from https://www.nato.int/cps/en/natohq/topics_50349.htm#:~:text=The%20PfP%20was%20established%20in,every%20field%20of%20NATO%20activity

NATO. (2022). *Extraordinary virtual summit of NATO Heads of State and Government*. Retrieved July 29, 2022, from https://www.nato.int/cps/en/natohq/news_192453.htm

Naylor, T. (2020). All that's lost: The hollowing of summit diplomacy in a socially distanced world. *The Hague Journal of Diplomacy*, *15*(2), 583–598. doi:10.1163/1871191X-BJA10041

NHSX. (2020). *Driving forward the digital transformation of health and social care*. NHSX. https://www.nhsx.nhs.uk/.

Nicolson, H. (1941). *Diplomacy*. Oxford University Press.

Nwachukwu, J. (2020, April 26). Nigerian lawyers drag China to court over COVID-19, demands $200b damages. *Daily Post*. https://dailypost.ng/2020/04/26/nigerian-lawyers-drag-china-to-court-over-covid-19-demands-200b-damages/

Nye, J. (2005). *Soft power: The means to success in the World politics*. Public Affairs.

O'Reilly-Shah, V. N., Gentry, K. R., Van Cleve, W., Kendale, S. M., Jabaley, C. S., & Long, D. R. (2020, May 12). The COVID-19 pandemic highlights shortcomings in US health care informatics infrastructure: A call to action. *Anesthesia and Analgesia*, *131*(2), 340–344. doi:10.1213/ANE.0000000000004945 PMID:32366769

Obiezu, T. (2020). Coronavirus concerns spur Nigerian authorities to close Chinese market in Abuja. *Voice of America*. https://www.voanews.com/science-health/coronavirus-outbreak/coronavirus-concerns-spur-nigerian-authorities-close-chinese

Obiorah, N. (2008). *Rise and right in China-Africa relations: (SAIS) Working paper in African series,* Washington: School of advanced International.

Okafor, L. (2020, March 17). Chinese responsible for direct importation of fake products to Nigeria. *Vanguard*. https://www.vanguardngr.com/2020/03/chinese-responsible-for-direct-importation-of-fake-products-to-nigeria-okafor/

Okeke, J. (2020, April 15). Onyeama attributes alleged ill-treatment of Nigerians in China on communication gaps. *Authority*. https://authorityngr.com/2020/04/15/onyeama-attributes-alleged-ill-treatment-of-nigerians-in-china-on-communication-gaps/

Olander, E. 2020. Nigeria's unprecedented censure of China. *The China-Africa Project*. https://chinaafricaproject.com/analysis/nigerias-unprecedented-censure-of-china/

Olusanya, G. (2012). 'Nigeria's role in the OAU'. *World event*, 2, pp.25-29.

Ong'ong'a, D. O. (2021). Systematic literature review: Online digital platforms utilisation by the ministry of foreign affairs in adopting digital diplomacy. *International Journal of Arts. Sciences and Humanities*, *9*(1), 8–18.

Oni, R. (2022). Nigeria Urged to Allow Use of China's COVID-19 Vaccines. *This day*. https://www.thisdaylive.com/index.php/2021/11/05/nigeria-urged-to-allow-use-of-chinas-covid-19-vaccines/

Onoja, A. (2020). How China lost Nigeria. *The Diplomat*. https://thediplomat.com/2020/08/how-china-lost-nigeria/

Onyeama, G. (2021, January 5). Nigerian Government Begins Talks With China Over COVID-19 Vaccines. *Sahara Reporters*. https://saharareporters.com/2021/01/05/nigerian-government-begins-talks-china-over-covid-19-vaccines

Orizu, U. 2020. COVID-19: Maltreatment of Nigerians in China resolved, says Gbajabiamila. *ThisDay*. https://www.thisdaylive.com/index.php/2020/04/15/covid-19-maltreatment-of-nigerians-in-china-resolved-says-gbajabiamila/

Osetskyi, V. (2019). IoT in healthcare: Use cases, trends, advantages and disadvantages. *Existek*. https://existek.com/blog/iot-in-healthcare/

Oshodi, A. G. T. (2020). Nigeria and China: Understanding the imbalanced relationship. *The African Report*. https://www.theafricareport.com/29060/nigeria-and-china-understanding-the-imbalanced-relationship/

Ouassini, A., Amini, M., & Ouassini, N. (2022). #ChinaMustexplain: Global tweets, covid-19, and anti-black racism in China. *The Review of Black Political Economy*, *49*(1), 61–76. https://doi.org/10.1177/0034644621992687

Page, M. T. (2018). *The intersection of China's commercial interests and Nigeria's conflict landscape*. Special Report by the United States Institute of Peace.

Pearce, K. (2020). Pandemic simulation exercise spotlights massive preparedness gap. *HUB*. https://hub.jhu.edu/2019/11/06/event-201-healthsecurity/

Perthes, V. (2021). Dimensions of rivalry: China, the United States, and Europe. *China International Strategy Review*, *3*, 56–65. doi:10.1007/s42533-021-00065-z

Peru T. (2022). Digital diplomacy is the next normal. *Diplomatic Courier*, 20-22

Pingjian, Z. (2020a). *NCP: Time for facts, science and solidarity*. Embassy of the People's Republic of China in the Federal Republic of Nigeria. http://ng.china-embassy.gov.cn/eng/zngx/cne/202002/t20200211_7775252.htm

Pingjian, Z. (2020b). *Ambassador Zhou Pingjian: China's response on COVID-19 responsible, decisive, effective*. Embassy of the People's Republic of China in the Federal Republic of Nigeria. http://ng.china-embassy.gov.cn/eng/zngx/cne/202002/t20200217_7775259.htm

Pingjian, Z. (2020c). *Ambassador Zhou Pingjian: Working together towards a health silk road*. Embassy of the People's Republic of China in the Federal Republic of Nigeria. http://ng.china-embassy.gov.cn/eng/zngx/cne/202003/t20200313_7775276.htm

Pingjian, Z. (2020d). COVID-19: *How China did it*. Embassy of the People's Republic of China in the Federal Republic of Nigeria. http://ng.china-embassy.gov.cn/eng/zngx/cne/202004/t20200403_7775281.htm

Pingjian, Z. (2020e). *Opening Remarks of Ambassador Zhou Pingjian of China at the Embassy's media briefing on fighting 2019 n-CoV*. Embassy of the People's Republic of China in the Federal Republic of Nigeria. http://ng.china-embassy.gov.cn/eng/zngx/cne/202002/t20200204_7775249. htm

Pingjian, Z. (2020f). *COVID-19: Hardship reveals true friendship*. Embassy of the People's Republic of China in the Federal Republic of Nigeria. http://ng.china-embassy.gov.cn/eng/zngx/cne/202002/t20200229_7775268.htm

Pingjian, Z. (2020g). *COVID-19: Economic consequence on China's economy transitory, manageable*. Embassy of the People's Republic of China in the Federal Republic of Nigeria. http://ng.china-embassy.gov.cn/eng/zngx/cne/202003/t20200306_7775272.htm

Pingjian, Z. (2020h). *Remarks of Ambassador Zhou Pingjian at the dialogue held by Centre for China Studies in Nigeria on the theme of "Nigeria-China Cooperation in the context of health emergency: Imperative of joint efforts and collaboration"*. Embassy of the People's Republic of China in the Federal Republic of Nigeria. http://ng.china-embassy.gov.cn/eng/zngx/cne/202002/t20200213_7775255.htm

Pingjian, Z. (2020i). *The Knowledge Center for China's Experiences in response to COVID-19*. Embassy of the People's Republic of China in the Federal Republic of Nigeria. http://ng.china-embassy.gov.cn/eng/zngx/cne/202003/t20200319_7775278.htm

Pingjian, Z. (2020j). *Statement by Press Secretary of the Embassy of China in Nigeria*. Embassy of the People's Republic of China in the Federal Republic of Nigeria. http://ng.china-embassy. gov.cn/eng/zngx/cne/202004/t20200412_7775289.htm

Pingjian, Z. (2020k). *Pandemic: solidarity and cooperation most potent weapon*. Embassy of the People's Republic of China in the Federal Republic of Nigeria. http://ng.china-embassy.gov. cn/eng/zngx/cne/202005/t20200512_7775336.htm

Pingjian, Z. (2020l). *Fighting Covid-19 to build a global community of health for all*. Embassy of the People's Republic of China in the Federal Republic of Nigeria. http://ng.china-embassy. gov.cn/eng/zngx/cne/202006/t20200613_7775360.htm

Pingjian, Z. (2020m). *The strength of China-Africa solidarity in defeating COVID-19*. Embassy of the People's Republic of China in the Federal Republic of Nigeria. http://ng.china-embassy. gov.cn/eng/zngx/cne/202006/t20200625_7775385.htm

Plessis, A. (2008). Defense diplomacy: Conceptual and practical dimensions with specific reference to South Africa. *Strategic Review for Southern Africa*, *30*(2), 87–119.

Power, M., & Mohan, G. (2010). Towards a critical geopolitics of China's engagement with African development. *Geopolitics*, *15*(3), 462–495. doi:10.1080/14650040903501021

Purwasito, A., & Kartinawati, E. (2020). *Hybrid Space and Digital Diplomacy in Global Pandemic Covid*, *19*. doi:10.2991/assehr.k.201219.100

Quin, D. (2020). Nigerian living near a major Belt and Road project *Pew Review*. https://www.pewresearch.org/fact-tank/2020/04/23/nigerians-living-near-a-major-belt-and-road-project-grew-more-positive-toward-china-after-it-was-completed/

Radwan, E., & Radwan, A. (2020). The-spread-of-the-pandemic-of-social-media-panic-during-the-covid-19-outbreak. *European Journal of Environment and Public Health*, *4*(2), em0044. doi:10.29333/ejeph/8277

Raj, S., Abu-Ghname, A., Davis, M. J., & Maricevich, R. S. (2020, September 1). The COVID-19 Pandemic: Implications for Medical Students and Plastic Surgery Residency Applicants. *Plastic and Reconstructive Surgery*, *146*(3), 396e–397e. doi:10.1097/PRS.0000000000007232 PMID:32541533

Report, A. W. (2021). *Drone technology propels Turkey's defense diplomacy and exports.* Retrieved December 9, 2021, from https://aviationweek.com/aerospace/emerging-technologies/drone-technology-propels-turkeys-defense-diplomacy-exports

Resdal. (1994). *Germany White Paper 1994.* Retrieved July 27, 2022, from https://www.resdal.org/Archivo/d0000066.htm

Rick, J. (2020). *3 reasons why civil society is essential to COVID-19 recovery.* WEF. https://www.weforum.org/agenda/2020/05/why-civil-society-is-essential-to-covid-19-pandemic-recovery/

Ritto, L. (2014, October 18). Diplomacy and its practice vs digital diplomacy. *Diplomat Magazine*. http://www.diplomatmagazine.nl.diplomacy-practice-vs-digital-diplomacy-2.

Roberts, W. R. (2007). What is public diplomacy? Past practices, present conduct, possible future. *Mediterranean Quarterly*, *18*(4), 36–52. doi:10.1215/10474552-2007-025

Rogers, R. (2021). Internet of Things-based Smart Healthcare Systems, Wireless Connected Devices, and Body Sensor Networks in COVID-19 Remote Patient Monitoring. *American Journal of Medical Research (New York, N.Y.)*, *8*(1), 71–80. doi:10.22381/ajmr8120217

Russell, W., Austin, G. L., Barton, K., Nugent, N., Kerr, D. S., Neil, R. S., & Lee-Lawrence, T. A. (2021). *'Quality' Response to COVID-19: The Team Experience of the Office of Quality Assurance, University of Technology*. CCEAM.

S., P. K., Sampath, B., R., S. K., Babu, B. H., & N., A. (2022). Hydroponics, Aeroponics, and Aquaponics Technologies in Modern Agricultural Cultivation. In *Trends, Paradigms, and Advances in Mechatronics Engineering* (pp. 223–241). IGI Global. https://doi.org/ doi:10.4018/978-1-6684-5887-7.ch012

Sabbati G, Tkalec I. Living in the EU: Work before the coronavirus crisis.

Sachs, J. D., Karim, S. A., Aknin, L., & Allen, J. (2020). Lancet COVID-19 Commission Statement on the occasion of the 75th session of the UN General Assembly. *The Lancet*.

Saldi, S. A. (2018). 'Editorial'. In De la Rochefocould A and Saldi S. (eds). The humanisation of robots and the robotisation of humans. Ethical reflections on lethal authonomous weapon systems and augmented soldiers. Chambesy: The Caritras in Veritate Foundation.

Samikannu, R., Koshariya, A. K., Poornima, E., Ramesh, S., Kumar, A., & Boopathi, S. (2022). Sustainable Development in Modern Aquaponics Cultivation Systems Using IoT Technologies. In *Human Agro-Energy Optimization for Business and Industry* (pp. 105–127). IGI Global.

Sampath, B. C. S., & Myilsamy, S. (2022). Application of TOPSIS Optimization Technique in the Micro-Machining Process. In Trends, Paradigms, and Advances in Mechatronics Engineering (pp. 162–187). IGI Global. doi:10.4018/978-1-6684-5887-7.ch009

Sandre, A. (2020). In review: top 10 moments in digital diplomacy. *The Medium*. https://medium.com/digital-diplomacy/2020-in-review-top-10-moments-in-digital-diplomacy-57b802e0159c

Sautman, B. (1994). Anti-black racism in Post-Mao China. *The China Quarterly*, *138*, 413–437. doi:10.1017/S0305741000035827

Schwab, K. (2016). The fourth industrial revolution. *Geneva: World Economic Forum*. Springer.

Seum, S. (2020). Guangzhou: Facts, solidarity and cooperation. Leadership.

Shankari, S., Rani, L., Brundha, & Somasundaram, J. (2020). Knowledge and Awareness on Role of Social Media in Managing COVID-19 Among General Population- A Questionnaire Study. *International Journal of Current Research and Review*, *12*(19), 197–202. doi:10.31782/IJCRR.2020.SP25

Shaum, R. (2016). Executive Summary: Digital diplomacy 20: Beyond the social media obsession. In S. Riordan (Ed.), *The strategic use of digital and public diplomacy in pursuit of national objectives* (pp. 2–4). Forcir Pensament.

Shenhav, S. R., Sheafer, T., & Gabay, I. (2010). Incoherent narrator: Israeli public diplomacy during the disengagement and the elections in Palestinian authority. *Israel Studies*, *15*(3), 143–162. doi:10.2979/isr.2010.15.3.143

Sherry, S. T., Ward, M. H., Kholodov, M., Baker, J., Phan, L., Smigielski, E. M., & Sirotkin, K. (2001, January 1). dbSNP: The NCBI database of genetic variation. *Nucleic Acids Research*, *29*(1), 308–311. doi:10.1093/nar/29.1.308 PMID:11125122

Shibayan, D. (2020, February 28). Senator asks FG to ban flights to and from China over coronavirus. *The Cable*. https://www.thecable.ng/senator-asks-fg-to-ban-flights-to-and-from-china-over-coronavirus

Shih, C. (2013). Harmonious Racism: China's Civilizational Soft Power in Africa. *Sinicizing International Relations*. Palgrave Macmillan. doi:10.1057/9781137289452_3

Shumba, E. (2020, Ocober 09). Twiplomacy in Africa: Possibilities and pitfalls for diplomats. *Africa portal*. https://www.africaportal.org/features/twiplomacy-africa-possibilities-and-pitfalls diplomats/

Shwab, K. (2019). *The fourth industrial revolution: What it means. How to respond.* World Economic Forum. https://www.weforum.org/agenda/2016/01/the-fourth-industrial-revolution-what-it-means-and-how-to-respond/

Singh, J. (2021). Military diplomacy: An appraisal in the Indian context. *CLAWS Journal, 15*(2), 108–124.

Spies, K. Y. (2018). African diplomacy. In G. Martel (Ed.), *Encyclopaedia of diplomacy* (pp. 1–14). John Wiley and Sons. doi:10.1002/9781118885154.dipl00C5

Sputnik. (2022). *ABD ile G. Kore, online askeri tatbikatla damarına bastıkları K. Kore'ye 'masaya dön' çağrısı yaptı.* Retrieved July 29, 2022, from https://tr.sputniknews.com/20220418/abd-ile-g-kore-dijital-askeri-tatbikatla-damarina-bastiklari-k-koreye-masaya-don-cagrisi-yapti-1055656984.html

Sudworth, J. (2021). Wuhan marks its anniversary with triumph and denial. *BBC.* https://www.bbc.co.uk/news/world-asia-china-55765875

Tara, N. (2022). *Africa needs smarter investment in digital infrastructure: Strategies for enticing the private sector. Foresight Africa: Technological innovations: Creating and harnessing tools for improved livelihood.* Foresight Africa.

Tavakoli, M., Carriere, J., & Torabi, A. (2020, July). Robotics, smart wearable technologies, and autonomous intelligent systems for healthcare during the COVID-19 pandemic: An analysis of the state of the art and future vision. *Advanced Intelligent Systems., 2*(7), 2000071. doi:10.1002/aisy.202000071

Taylor, I. (2007). China's relations with Nigeria. *The Round Table, 96*(392), 631–645. doi:10.1080/00358530701626073

Tettey, J. (n.d.). Anti-black racism in China, and political economy of asymmetrical power. *Intervention Symposium – "Black Humanity: Bearing Witness to COVID-19" organized by Elaine Coburn and Wesley Crichlow COVID-19.* https://antipodeonline.org/wp-content/uploads/2020/12/7.-Tettey.pdf

Tham, D. (2020, January 09). Taiwan's digital diplomacy gets a kickstart. *Taipei Times.* https://www.taipeitimes.com/News/feat/archives/2020/01/09/2003728948

Thinktech, S. T. M. (2022). *Yeni bir paradigm: Askeri metaverse ve geleceği.* Retrieved July 29, 2022, from https://thinktech.stm.com.tr/tr/yeni-bir-paradigma-askeri-metaverse-ve-gelecegi

ThisDay. (2015). China–Nigeria ties as a framework for attaining UN Sustainable Development Goals. *This Day.* http://allafrica.com/stories/201511152091.html

Thorpe, J. (2010). *Trends in modeling, simulating & gaming: Personel observations about the past thirty years and speculation about the next ten.* Retrieved July 29, 2022, from https://www.iitsec.org/-/media/sites/iitsec/link-attachments/iitsec-fellows/2010fellows_thorpe.ashx?la=en

Tse, D. K., & Hung, K. (2014). *Chinese firms going global: Their impacts, best practices, and implications*. Cambridge University Press.

Turchetti, S., & Lalli, R. (2020). Envisioning a "science diplomacy 2.0": On data, global challenges, and multi-layered networks. *Humanit Soc Sci Commun, 7*(1), 144. doi:10.105741599-020-00636-2

Turianskyi, Y., & Wekesa, B. (2021). African digital diplomacy: Emergence, evolution, and the future. *South African Journal of International Affairs, 28*(3), 341–359. doi:10.1080/10220461.2021.1954546

Türkiye Cumhuriyeti Dışişleri Bakanlığı. (2022). *Türkiye Cumhuriyeti Dışişleri bakanlığı tarihçesi*. Retrieved August 15, 2022, from https://www.mfa.gov.tr/turkiye-cumhuriyeti-disisleri-bakanligi-tarihcesi.tr.mfa#:~:text=1924%20y%C4%B1l%C4%B1nda%2039%20d%C4%B1%C5%9F%20temsilcili%C4%9Fe,toplam%20253%20misyona%20sahip%20bulunmaktad%C4%B1r

Umejei, E. (2015). China's engagement with Nigeria: Opportunity or opportunist? *Africa East-Asian Affairs, 3*(4), 54–78. doi:10.7552/0-3-4-165

United Nation Organisation. (2021). *Digital diplomacy in the era of COVID-19*. UNO.

Unver, A. (2017). *Computational diplomacy*. EDAM.

Van der Linden, S. (2015). The conspiracy-effect: Exposure to conspiracy theories (about global warming) decreases pro-social behaviour and science acceptance. *Personality and Individual Differences, 87*, 171–173.

Vanitha, S. K. R., & Boopathi, S. (2023). Artificial Intelligence Techniques in Water Purification and Utilization. In *Human Agro-Energy Optimization for Business and Industry* (pp. 202–218). IGI Global. doi:10.4018/978-1-6684-4118-3.ch010

Vargo, D., Zhu, L., Benwell, B., & Yan, Z. (2020). Digital technology use during -19 pandemic: A rapid review. *Human Behavior and Emerging Technologies*.

Vennila, T., Karuna, M. S., Srivastava, B. K., Venugopal, J., Surakasi, R., & Sampath, B. (2022). New Strategies in Treatment and Enzymatic Processes: Ethanol Production From Sugarcane Bagasse. In *Human Agro-Energy Optimization for Business and Industry* (pp. 219–240). IGI Global.

Verma, R. (2020). China's diplomacy and changing the COVID-19 narrative. *International Journal (Toronto, Ont.), 75*(2), 248–258. doi:10.1177/0020702020930054

Vincent, D. (2020, April 17). Africans in China: We face coronavirus discrimination. *BBC News*. (2020). https://www.bbc.com/news/world-africa-52309414

Vincent, J. (2020, June 3). Something in the air: Conspiracy theorists say 5G causes novel coronavirus, so now they're harassing and attacking UK telecoms engineers. *The Verge*. https://www.theverge.com/2020/6/3/21276912/5g-conspiracy-theories-coronavirus-uk-telecoms-engineers-attacks-abuse

Virani, S. S., Maddox, T. M., Chan, P. S., Tang, F., Akeroyd, J. M., Risch, S. A., Oetgen, W. J., Deswal, A., Bozkurt, B., Ballantyne, C. M., & Petersen, L. A. (2015, October 20). Provider type and quality of outpatient cardiovascular disease care: Insights from the NCDR PINNACLE Registry. *Journal of the American College of Cardiology*, *66*(16), 1803–1812. doi:10.1016/j.jacc.2015.08.017 PMID:26483105

Volodenkov, S., & Pastarmadzhieva, D. (2020). Digital society in the context of the COVID-19 pandemic: First results and prospects (comparative analysis of the experience of Russia and Bulgaria). *Journal of Political Research.*, *4*(2), 80–89. doi:10.12737/2587-6295-2020-80-89

Waithaka, I. N. (2018). *Digital diplomacy: The integration of information and communication technologies in Kenya's Ministry of foreign Affairs 1964-2014.* [MA Thesis, University of Kenyatta University].

Wekesa, B. (2020, July 03). Pathways for theorising African digital diplomacy. *Africa Portal.* https://www.africaportal.org/features/pathways-theorising-african-digital-diplomacy/

Wekesa, B., Turianskyi, Y., & Ayodele, O. (2021). Introduction to the special issue: Digital diplomacy in Africa. *South African Journal of International Affairs*, *28*(3), 335–339. doi:10.1080/10220461.2021.1961606

White, H. (2014). Grand expectations, little promise. In B. Taylor, J. Blaxland, H. White, N. Bisley, P. Leahy, & S. S. Tan (Eds.), *Defense Diplomacy Is the game worth the candle?* (pp. 10–11). ANU Strategic and Defense Studies Centre.

Whitelaw, S., Mamas, A., Topol, E., Van Spall, H. (2020). Applications of digital technology in COVID-19 pandemic planning and response. *The Lancet Digital Health*.

Winger, G. (2014). The velvet gauntlet: A theory of defense diplomacy. In What Do Ideas Do IWM Junior Visiting Fellows' Conferences (vol. 33, pp. 1-15). Vienna: IWM.

Wingo, A. (2017). Akan philosophy of the person. *Stanford Encyclopaedia of Philosophy* (1-14), Stanford: Stanford University Press.

Wong, C. A., Ming, D., Maslow, G., & Gifford, E. J. (2020). Mitigating the Impacts of the COVID-19 Pandemic Response on At-Risk Children. *Pediatrics*, *146*(1), e20200973. doi:10.1542/peds.2020-0973 PMID:32317311

World Health Organisation. (2020 February 15). Munich Security Conference. WHO. https://www.who.int/dg/speeches/detail/munich-security-conference

World Health Organisation. (2020). *Origin of SARS-CoV-2*. WHO. https://apps.who.int/iris/bitstream/handle/10665/332197/WHO-2019-nCoV-FAQ-Virus_origin-2020.1-eng.pdf

Yackley, A. J. (2020). *How Turkey militarized its foreign policy*. Retrieved October 15, 2020, from https://www.politico.eu/article/how-turkey-militarized-foreign-policy-azerbaijan-diplomacy/

Zaharma, R. S. (2018). *Battles to bridges: US strategic communication and public diplomacy after 9/11*. Springer.

Zeng, J., Mike, S., & Schafer, S. (2021). Conceptualising "Dark Platforms". Covid-19-related conspiracy theories on 8kun and Gab. *Digital Journalism*, *41*(2), 1–23.

Zhou, P. (2020a). China's response on COVID-19 responsible, decisive, effective. Leadership, January 29, edition, p.44.

Zhou, P. (2020b). COVID-19: Hardship reveals true friendship. People Daily, (February), p.22

Zhou, P. (2020c). COVID-19: How China did it. Leadership, (April), pp. 22

Zhou, P. (2020d). NCP: Time for facts, science and solidarity. New Telegraph, (February), p.30.

Zhu, A. (2020). A lost 'little Africa': How China, too, blames foreigners for the virus. *The New York Review*. https://www.nybooks.com/online/2020/05/05/a-lost-little-africa-how-china-too-blames-foreigners-for-the-virus/

Zsubrinzky, Z. (2020). Digital communication in diplomacy. Paper presented in the *12*th *International Conference of J. Selye University Pedagogical Sections,* (pp, 147-155). J. Selye University.

About the Contributors

Floribert Endong (PhD) is a research consultant in the humanities and social sciences. He is a reviewer and editor with many scientific journals in the social sciences. His current research interest focuses on advertising, cinema and globalization, international communication, gender studies, digital media, media laws, international relations, culture and religious communication. He is author of numerous peer-reviewed articles and book chapters in the above mentioned areas of interest. He recently edited a book titled *Exploring the Role of Social Media in Transnational Advocacy*, published by IGI Global.

* * *

M. Aaschita Reddy works within the Department of Electronics and Communication Engineering, Sreenidhi Institute of Science and Technology, Hyderabad, Telangana, India.

Ekrem Yasar Akçay is an associate professor at Hakkâri University, Department of Political Science and International Relations. He completed his master studies at Suleyman Demirel University, Department of International Relations in 2010. Then he got his PhD at Ankara University, Institute of Social Sciences, Department of International Relations. He was a researcher at Humboldt Universitat zu Berlin and Wroclaw University for six months. He has many articles that is published in numerous academic journals. His book "Türkiye-AB İlişkileri ve İmtiyazlı Ortaklık Meselesi" (The Privileged Partnership Question in Turkey-EU Relations) is published in February 2016 by Imaj Publishers. His other book, "Soğuk Savaş Döneminde Avrupa: Temel Konular Üzerinden Avrupa'nın Dönüşümünü Anlamak" (Europe in the Cold War Era: Understanding the Transformation of Europe over the Issues) is published in September 2017 by Kriter Publishing.

Chinonso Aniagu, PhD, is a public relations consultant who specializes in advancing postmodernist approaches to communication-oriented research. He holds a PhD in Media Studies from the University of Calabar and is a member of Nigerian Institute of Public Relations (NIPR).

Sampath Boopathi () completed his undergraduate in Mechanical Engineering and postgraduate in the field of Computer-Aided Design. He completed his Ph.D. from Anna University and his field of research includes Manufacturing and optimization. He published 60 more research articles in Internationally Peer-reviewed journals, one Patent grant, and three published patents.He has 16 more years of academic and research experiences in the various Engineering Colleges in Tamilnadu, India..

Grace Eugenie Essoh holds a PhD in Spanish linguistics from the University of Dschang in Cameroon. She teaches foreign languages at the Department of Modern Languages and Translation Studies of the University of Calabar, Nigeria. Her areas of interest include diplomatic communication, discourse analysis, anthropological linguistics and cultural studies. She has published numerous peer-reviewed journal articles and book chapters in the above-mentioned areas of interest.

Ranson Sifiso Gwala holds a Doctor of Business Administration (DBA) from the University of KwaZulu Natal (UKZN). He has published more than 10 peer-reviewed articles. He also attended and presented papers at the international conferences starting in 2022. His interests are corporate governance, marketing, leadership, and leadership development in both the public and private sectors. He believes that academia should be an integral part of supporting the public sector to deliver services and improve the lives of people. Academia is a at a vantage point to provide meaningful developmental imperatives for all sectors of society. He was born in Ndwedwe, KwaZulu-Natal, but now resides between Durban and Pietermaritzburg, South Africa.

Fulufhelo Makananise is an Associate Professor in Media Studies in the Department of Communication Science at the University of South Africa (UNISA). He holds a Doctoral degree in Media Studies from the University of Limpopo and has obtained a PG (dip)HE from Rhodes University. He serves as an external examiner in other South African universities and acts as a reviewer in both international and national scholarly journals and book projects. Professor Makananise has authored and published academic articles in peer-reviewed and DHET-accredited journals. He has authored several chapters published in peer-reviewed books. In addition, Prof Makananise has presented papers both at national and international conferences. His research interest is in new media technology, digital media, indigenous language media, social media, political communication. digital diplomacy, and news media consumption.

B Manidweep Reddy is involved with H.No. 4-35, Opposite Sadashiva High School, Ferozguda, Tirumalagiri, Hyderabad,Telangana, India.

Pfano Mashau is an academic at the University of KwaZulu-Natal, School of Entrepreneurship and Management. He holds a PhD in Management. He also has various non-degree career development certificates acquired from short programmes (Locally and Internationally). As an academic, he is involved in lecturing at the undergraduate and postgraduate levels. His research focus is on Small Business Development, Innovation, Business Management, Entrepreneurship and Agglomeration Economies. Prior to working as an academic, he worked for JET Education Services, BioRegional and a few small businesses. Prof Mashau has published over 40 research articles, supervised eight doctoral theses to completion, and over 20 Masters dissertations. Some of the research work has been presented in international conferences. He has been an editor for a journal and a book. He is a member of the Pan African Research Council and the Institute of Business Advisors. He was briefly appointed as eThekwini Municipality City Planning Commissioner. Prof Mashau aspires to see researchers conducting impactful and life-changing research studies. Prof Mashau interests outside work are cycling, skateboarding, and jogging.

C.S. Mukund works in the department of Electronics and Communication Engineering, Sreenidhi Institute of Science and Technology, Hyderabad, Telangana, India.

Murat Mutlu got his MA degree at Hakkari University, Department of Political Sciences and International Relations.

D. M. D. Preethi works in the Department of Computer Science and Engineering, PSNA College of Engineering and Technology (Autonomous), Dindigul, Tamil Nadu, India.

Ndivhuwo Sundani is a Lecturer at the University of South Africa's Department of Communication Science. He has published four journal articles and presented six papers at academic conferences and research seminars to date.

Kiran Venneti is working as Assistant professor in the Department of Electronics and Communication Engineering, Aditya College of Engineering, Surampalem, Andra Pradesh, India.

Index

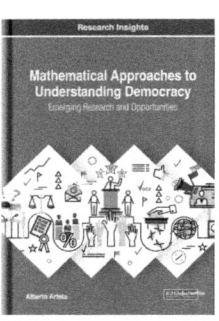

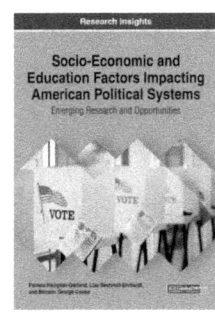

Are You Ready to
Publish Your Research ?

IGI Global
PUBLISHER of TIMELY KNOWLEDGE

IGI Global offers book authorship and editorship opportunities across 11 subject areas, including business, computer science, education, science and engineering, social sciences, and more!

Benefits of Publishing with IGI Global:

- Free one-on-one editorial and promotional support.

- Expedited publishing timelines that can take your book from start to finish in less than one (1) year.

- Choose from a variety of formats, including Edited and Authored References, Handbooks of Research, Encyclopedias, and Research Insights.

- Utilize IGI Global's eEditorial Discovery® submission system in support of conducting the submission and double-blind peer review process.

- IGI Global maintains a strict adherence to ethical practices due in part to our full membership with the Committee on Publication Ethics (COPE).

- Indexing potential in prestigious indices such as Scopus®, Web of Science™, PsycINFO®, and ERIC – Education Resources Information Center.

- Ability to connect your ORCID iD to your IGI Global publications.

- Earn honorariums and royalties on your full book publications as well as complimentary content and exclusive discounts.

Join Your Colleagues from Prestigious Institutions, Including:

Australian National University

Massachusetts Institute of Technology

JOHNS HOPKINS UNIVERSITY

HARVARD UNIVERSITY

COLUMBIA UNIVERSITY IN THE CITY OF NEW YORK

Learn More at: www.igi-global.com/publish
or by Contacting the Acquisitions Department at: acquisition@igi-global.com

Ingram Content Group UK Ltd.
Milton Keynes UK
UKHW050149110523
421555UK00007B/62